Springer-Verlag Berlin Heidelberg GmbH

Martin J. Tovée

The Speed of Thought:

Information Processing in the Cerebral Cortex

Springer

Martin J. Tovée

Newcastle University
Newcastle upon Tyne
United Kingdom

ISBN 978-3-662-10410-1 ISBN 978-3-662-10408-8 (eBook)
DOI 10.1007/978-3-662-10408-8

Biotechnology Intelligence Unit

Library of Congress Cataloging-in-Publication data

Tovée, Martin J. (Martin James)
The speed of thought: information processing in the cerebral cortex / Martin J. Tovée.
p. cm. — (Neuroscience intelligence unit)
Includes bibliographical references and index.

1. Visual cortex. 2. Visual pathways. I. Title. II. Series.
[DNLM: 1. Visual Perception—physiology. 2. Cerebral Cortex—physioology. 3. Mental processes—physiology. WL 105 T736s 1998]
QP383.15.T68 1998
612.8'25—dc21
DNLM/DLC 98-11141
for Library of Congress CIP

Originally published by Springer-Verlag Berlin Heidelberg and R.G. Landes Company in 1998
Softcover reprint of the hardcover 1st edition 1998

Typesetting: R.G. Landes Company, Georgetown, TX, U.S.A.

SPIN 10685454 31/3111 - 5 4 3 2 1 0 - Printed on acid-free paper

Dedication

This book is dedicated to my wife Esther, and to our daughter Charlotte.

Acknowledgments

I have worked on the topics covered in this book with Edmund Rolls, Alessandro Treves, Larry Abbott, Vilayanur Ramachandran, Stefano Panzeri and Ray Bellis. I am grateful for discussion of the issues covered in this book with Malcolm Young, Jack Scannell, Huw Golledge, Claus Hilgetag, Gully Burns, Roger Mason, Dave Perrett, Mike Oram, and Hugo Critchley. Lastly, and most importantly, I would like to thank my wife and research collaborator Esther, for her critical reading of, and comments on, the manuscript.

CONTENTS

Chapter 1

The Visual System

The Need for Speed

The "speed of thought" may be measured in a number of ways by different disciplines (such as by cognitive psychologists and neuropsychologists). However, only the neurophysiologist is able to go to the heart of the matter; the speed of the response of neurons. Such investigations provide crucial information for all researchers working on how the brain works, as it allows them to base their ideas in a biologically plausible framework. However, almost all approaches to interpreting a neuron's response examine long time samples of the response, usually around 300-500 msec. Is this a biologically plausible length of time? We know that it is possible to recognize and respond to a visual stimulus within 400-600 msec. If one considers that at least half of this time is involved in the generation and implementation of motor commands, it means that the total amount of time available for processing in the whole visual system is considerably less than the period over which a cell's response is conventionally measured. This book focuses on the relatively neglected issue of the importance of rapid processing, and how we should take this into account as we explore how information is processed in the visual system.

This "need for speed" shapes the functional architecture of all sensory systems, acting to reduce, where possible, wiring length and the number of synapses, to allow faster processing. It is a primary constraint on organization that the nervous system must react rapidly to changes in the environment. This is particularly true of the visual system, which is the primary sensory modality of the primates, and must encode the most rapidly changing source of information. This need for rapid processing has important implications for recognizing and storing visual images. How information is encoded and decoded, the size of neural population in which information is

The Speed of Thought: Information Processing in the Cerebral Cortex,
by Martin J. Tovée.

encoded and how information distributed through the visual system is bound into a coherent percept are all affected and shaped by the time constraints in which these computations must be performed. In this book I will discuss the evidence of how information is processed and represented and then discuss how this has been affected by temporal considerations.

This book deals primarily with information processing in the primate temporal visual cortex, one of the higher visual association areas (Fig. 1.1). This area is believed to be important for the representation of complex stimuli and may also play a role in visual memory. However, the way in which these stimuli are encoded by temporal visual neurons can be used as a general model of visual and sensory processing. For example, the possibility that information is encoded in the temporal pattern of neural firing is a question that can be asked of the responses of any neuron in the brain, and the same analysis techniques are used to analyze spike trains whatever their origin. The general principles which can be drawn from studies of the temporal visual cortex can be applied to the functioning of any cortical area. However, before discussion of this area in detail, it is necessary to place it in its anatomical and functional context. Thus, in this chapter, I will discuss the cortical structure of the primate brain and the functional organization of the visual system.

The Cerebral Cortex

The mammalian cortex is a strip of neurons, usually divided into six layers. It varies in thickness from about 1.5 to 4.5 mm in humans, and this is not very different even for the very small cerebral hemispheres of the rat, where the thickness is about 1-2 mm thick. The most conspicuous difference is that the surface area increases enormously in higher animals. For example, the surface area ranges from 3-5 cm^2 per hemisphere in small brained rodents to 1,100 cm^2 in humans. It seems that the evolutionary expansion of the cortex may be constrained by the way the cortex is formed during development. A persuasive theory has been put forward by Pasko Rakic[1-2] based on the limitations that cell divisions during development place on the expansion of the cortex. This model, called the radial-unit hypothesis, proposes that the 1,000-fold increase in the expansion of cortical surface area seen in mammalian evolution is the result of changes in cell division that increase the number of radial columnar units which make up the cortex, without changing the number of cells in each column. Thus the sheet-like structure of the cortex is

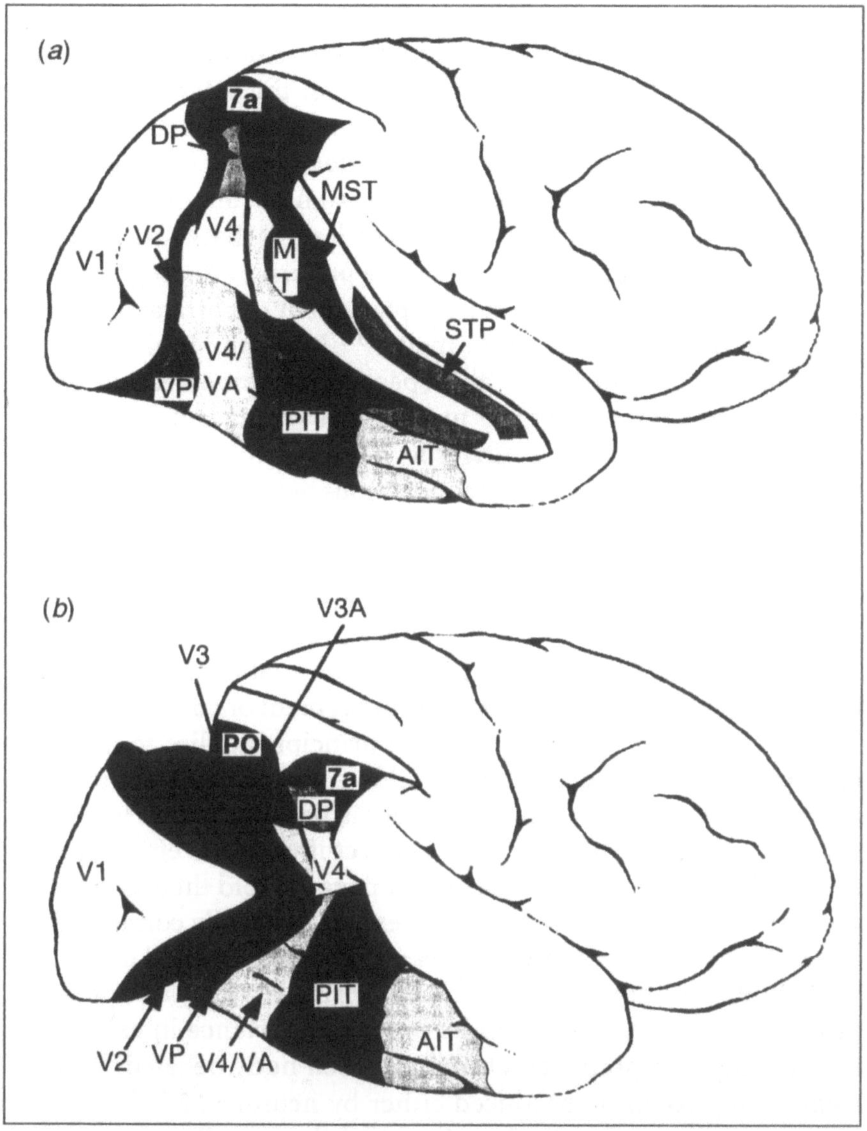

Fig. 1.1. The location of the major visual areas in the macaque cerebral cortex. (a) The superior temporal sulcus has been unfolded so that the visual areas normally hidden from view can be seen. (b) The lunate, inferior occipital and parieto-occipital sulci have been partially unfolded. AIT, anterior inferior temporal cortex; DP, dorsal prelunate; MT, middle temporal also called V5; MST, medial superior temporal; PIT, posterior inferior temporal cortex; PO, parieto-occipital; STP, superior temporal polysensory; VA, ventral anterior; VP, ventral posterior. (Redrawn from Maunsell JHR, Newsome WT. Ann Rev Neurosci 1987; 10:363-401.).

determined by the constraints of cell division during development. The cortical sheet is folded to produce a series of ridges (gyri), and furrows (sulci). The simplest explanation for this folding is that you have to fold a large sheet to get it into a small box. But mere folding explains neither the species specific pattern of sulci and gyri, nor why they provide landmarks to the location of functional areas of cortex, nor why their pattern is altered by lesions that cause the brain to 're-wire'.[1] So what factors control the pattern of folding?

One likely explanation for the placement of cortical folds is to minimize the volume of white matter.[3-5] It is commonly accepted that some, but by no means all, aspects of the organization of the central nervous system appear to minimize wiring volume.[6-8] Quite simply, an animal which arranges its neurons efficiently, by putting the computationally necessary connections between nearby neurons and leaving 'nonconnections' between neurons that are far apart, can make do with less white matter and will benefit from a smaller, faster and cheaper brain. Such a brain should also be easier to make with simple developmental and evolutionary processes.

Efficient wiring may be seen in neuronal arbors,[7-8] cortical maps,[6-7] and in the 2-dimensional arrangement of cortical areas.[9-11] There is also some evidence that the principle applies to the 3-dimensional morphology of cortical folds. A geometric description of the human cortex shows that the surface is 'peculiarly close together'[3] (Fig. 1.2), which is what one would expect if wiring is efficient. Furthermore, both the cat and macaque appear to fold their cortices in a way that devotes the available convexities to heavily connected areas and puts the concavities between sparsely connected areas.[4-5]

While the importance of efficient wiring is widely accepted, the processes that generate it and its overall importance in explaining major aspects of brain structure have been hotly debated.[12-13] Efficient wiring could be produced either by neurons and their territory starting in particular locations and then sending projections locally (local wiring), or by neurons and their territory starting out with particular connections and then "migrating" to get close to the things with which they connect (component placement optimization, CPO). The fact that wiring is efficient does not distinguish between these possibilities.

Until recently, developmental and evolutionary considerations suggested that local wiring rather than CPO could best account for the observed regularities between connectivity and location. Indeed,

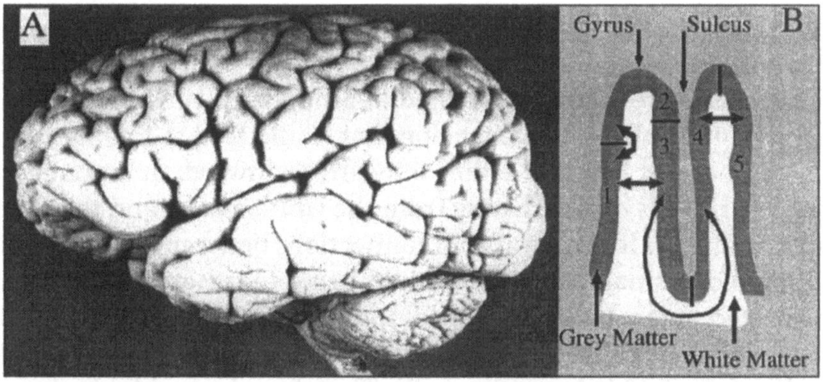

Fig. 1.2. (A) shows the human brain. In this and many other mammalian brains, a distinct pattern of folds is the most striking anatomical feature. The pattern is characteristic of species and is related to the mosaic of distinct functional areas which make up the cortex. (B) shows how folds may influence the length of corticocortical connections. In this model, 5 functional areas (areas 1 to 5) are distributed over 2 gyri. 1 and 2, and 3 and 4, are 'nearest neighbors' (NN), while 1 and 3, and 3 and 5 are 'next door but one' on the cortical sheet. Area 1 is 'nearest neighbor OR next door but one' (NDB1) with 2 and 3. Axons linking areas 1, 2 and 3 would be short, while axons linking 3 and 4 would be long. Thus, given the same axonal diameter, spike rate, and axon number, a corticocortical connection between 1 and 3 would be more compact, faster, and use less energy than a connection between 3 and 4. An efficiently folded cortex might place the folds so that heavily connected areas are together on gyri while sparsely connected areas are separated by sulci. (Figure courtesy of Dr. J.W. Scannell.)

the evidence that structures migrated around the brain to minimize wire is questionable.[13] However, when it comes to the 3D arrangement of cortical areas in relation to sulci and gyri, it does now look as if major brain structures may be positioned in a way that reduces connection length.[1]

Cortical Origami

The cortical sheet is a jigsaw of functionally distinct areas linked by a complex network of corticocortical connections. How is the folding coordinated with the wiring? Van Essen has suggested that two factors play a key role. The first are intrinsic factors, such as differential growth rates in the gray matter, and second are extrinsic factors, which are based on long-range axonal connections in the underlying white matter. Some of the axonal connections are to subcortical structures, and Van Essen hypothesizes that the tension generated in these axons produces an inward force to counteract the intraventricular hydrostatic force generated by the CSF. The second form of axonal connections are between different cortical areas. These connections are established at around the time that folding begins, and could generate tension that would lead to folding.

The cortex can fold either outward or inward. In an outward fold, the ridge is directed away from the white matter and the brain interior, and the length of axonal connections between the two banks of the fold is small. Such folds could bring together densely interconnected areas. In an inward fold, the crease is directed towards the white matter and so the white matter distance between the two banks of the fold is long (Fig. 1.2B). Therefore, inward folds should end up between sparsely connected areas. This suggestion is consistent with results recently published on connectivity and cerebral folding in the macaque and cat brain.[4] Heavily interconnected areas tend to be separated by gyri and sparsely connected areas seem to separated by sulci.

Thus one has to make a trade-off, reducing the tension in the axonal connections between some cortical areas at the price of increasing the tension in the connections between other areas. The connections between some areas are more extensive than those between other areas, so if one makes an outward fold at the boundary between two areas which are densely connected and an inward fold at the boundary between two sparsely connected areas, the overall axonal tension will be reduced. Thus, the eventual shape of the cortical sheet will be determined by the relative density of connections between different areas.

Other aspects of the gross morphology of the brain may follow from the same mechanisms. The link between wiring and folding is supported by evidence from developmental studies. For example, prenatal bilateral eye removal in the macaque alters the pattern of folding in the occipital cortex in the absence of direct mechanical intervention.[1] Thus, even if tension based factors do not turn out to be the explanation, developmental neuroscientists still need to account for the relationship between wiring and folding, possibly turning their attention to the possibility that growth factors are released by corticocortical axons.

While efficient wiring is an attractive principle, it should not blind us to the fact that brains represent a compromise between many competing constraints. As well as saving wire, brains have to produce adaptive behavior, they have to be made during development, specified by a genome, and based on a design inherited from the parents. It is unlikely that, in balancing all these constraints, the brain could be optimal for any one. Indeed, apparent examples of wire-wasting connectivity are widespread; the facts of developmental pruning, the inverted retina, the visual cortex at the wrong end of the brain, and the unconnected thalamic nuclei clustering together and not with the groups of cortical areas with which they exchange axons, all suggest that other factors are at work.[4,13]

Dividing Up the Brain

The brain is an incredibly complex structure. It contains around 10^{11} neurons, which have more than 10^{15} synapses and at least 2,000 miles of axonal connections.[13] Fortunately, for those of us who wish to make sense of how the brain works, there are several rules of organization which simplify our task. Firstly, neurons with similar patterns of connections and response properties are clustered together to form areas. For example, in the monkey and the cat there are about 70 cortical areas linked by around 1,000 connections. Connections between these brain areas may consist of tens of thousands or even millions of nerve fibers. Many of these areas seem specialized to perform different tasks, so, for example, visual area 5 (V5) seems specialized to process information on visual motion and visual area 4 (V4) seems at least partially specialized for color. The number of different specialized areas increases with increasing size and complexity of the brain. For example, mice have 15 cortical areas, of which around 5 are visual areas, whereas the cat has 65 cortical areas, of which 22 are visual.[11,15] It is suggested that the increase in visual areas allows the analysis of an increased number of visual parameters, which in turn allows a more complex and detailed analysis of visual stimuli. There is considerable interaction between neurons dealing with a particular visual parameter, such as color or motion, and by grouping all such neurons into specialized areas the amount and the length of connections between neurons is reduced. The arrangement and connections between neurons is largely predetermined genetically. To have widely interconnected neurons, and to have many different types of neurons with different connection patterns spread throughout the brain, would be extremely difficult to program genetically and would have a greater potential for errors.[15]

The Structure of the Visual System

Vision is the primary sensory modality in primates, and this is reflected in the complexity of the visual system and the extent of the cerebral cortex used for the analysis of visual information. It is believed that at least thirty-two separate cortical areas are involved with the processing of visual processing in the macaque monkey[16] (Fig. 1.1). Twenty-five of these areas are primarily visual in function; the remaining seven are also implicated in other functions such as polysensory integration or visually guided motor control. These visual areas occupy about half of the 100 cm^2 area of each of monkey's cortical hemispheres. Two of the areas, V1 and V2, each exceed more than 10 cm^2 of the cortical surface, but most visual areas occupy less than a tenth of this size. Comparatively little is known of the functional anatomy of the human visual cortex, but it seems to be at least as complex as that of the monkey.[17-18] Fortunately, it is possible to simplify this picture by concentrating on the key visual areas and looking at their functional organization. I will describe this system as a chain of feed-forward cortical areas, which is how it is commonly described. This is a permissible simplification as long as one never loses sight of the fact that it is not actually true. There are at least as many back-projections as forward, both within an area and between areas. Additionally, there are also considerable lateral connections between neurons and processing modules within an area, and between equivalent areas in different processing streams. The visual system is less a chain of areas leading forward, than a grid in which information simultaneously moves forward, sideways and back.

As one moves up the visual system, from the retina to the lateral geniculate nucleus (LGN) and then on to successive cortical areas, visual neurons become responsive to more and more complex stimuli. For example, in monkeys, in the first cortical visual area (called primary visual cortex or V1) there are neurons responsive to simple lines of different orientations, whereas in one of the higher visual areas (inferior temporal cortex) the neurons respond to complex stimuli, such as faces. However, the visual system is not organized in just a serial, hierarchical pathway. Different aspects of a stimulus (such as its shape, color and motion) are analyzed in separate, parallel pathways. These pathways are usually divided into two broad categories: "what" and "where" pathways. The "what" pathway deals with information about the stimulus features (such as shape and color) and the identity of an object, and can be subdivided into two further pathways: color and shape. The "where" pathway deals with spatial information about an object, and is usually subdivided into motion and form derived from motion.

The Origin of the M and P Pathways

This separation starts to become evident at the level of the retina. There are several forms of ganglion cell in the primate retina, of which two types (M and P cells) constitute about 90% of the cells. The M class (sometimes also called A or Pα cells) account for 10%, and the other 80% are accounted for by the P class (sometimes called B or Pβ cells). The subsequent pathways are often referred to as the M or P pathways (Fig. 1.3). The remaining 10% of the retinal ganglion cells consist of at least 8 different types. The P cells are selective for wavelength and high spatial frequencies and have tonic responses, whereas the M cells are not wavelength sensitive but are sensitive to low spatial frequencies, have phasic responses and have faster conduction velocities. At any one eccentricity, the dendritic field of the M cell is three times larger than that of the P cell. These differences in the response properties of the neurons shape the functions of the subsequent visual areas. The retinal ganglion cells project primarily to a six-layered structure called the dorsal lateral geniculate nucleus (LGN). The M ganglion cells project to cells in layers 1 and 2. Cells in these layers have relatively large cell bodies and so these layers are called the magnocellular (M) layers. The P ganglion cells project to layers 4-6. These have comparatively small cell bodies, and are so called parvocellular (P) layers. The LGN neurons show the same response characteristics as the cells from which they receive input.

The LGN neurons mainly project to the primary visual cortex (also known as the striate cortex or V1). This is the first cortical visual area and consists of six principal layers (and several sublayers) arranged in bands parallel to the surface of the cortex. The axons from the LGN terminate on cortical neurons in layer 4. The P layer neurons send their axons to neurons in the deeper part of this layer (sublayer 4Cβ), which in turn send their axons to layers 2 and 3 and from there to Visual Area 2 (V2). The M layer neurons send their axons to neurons to sublayer 4Cα, and the information is then relayed to layer 4B and then to V2 and to Visual Area 5 (V5). Cells in layer 4B are orientation selective and most show selectivity for the

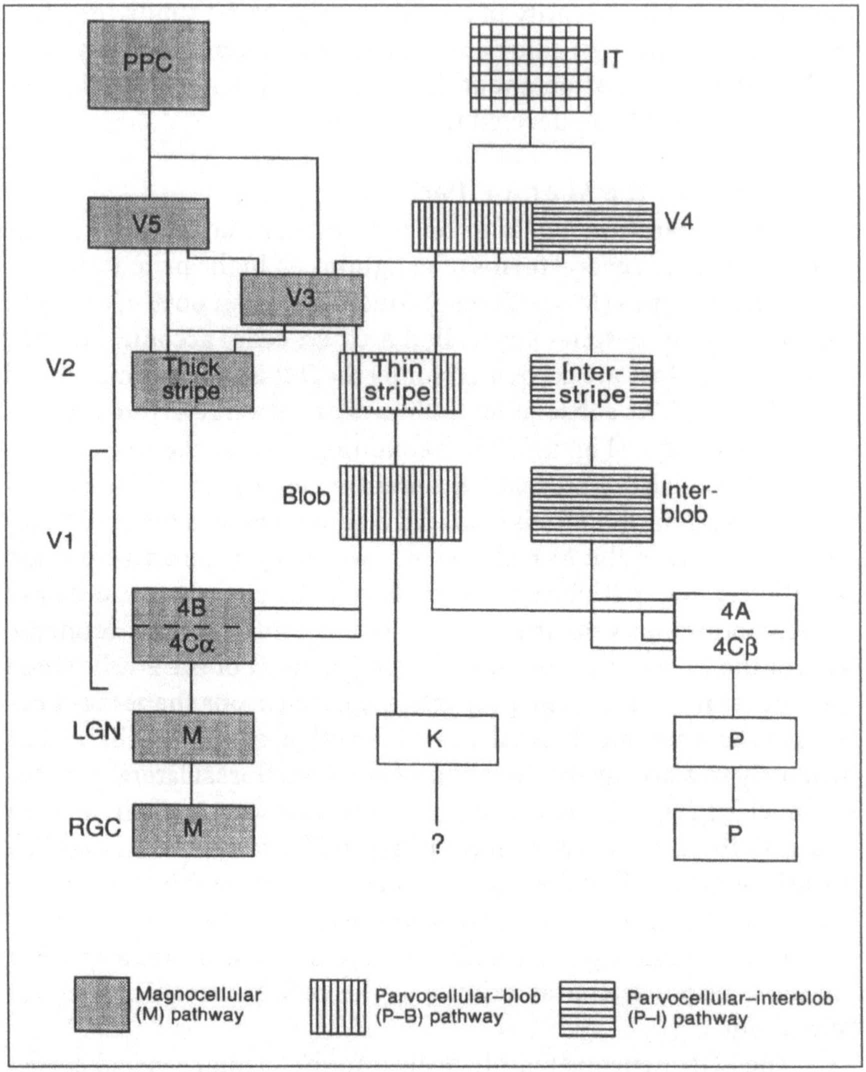

Fig. 1.3. Subcortical and cortical pathways in the macaque. There are two main pathways; the parvocellular (P) pathway and magnocellular (M) pathway. The P pathway splits to produce two new pathways in the upper layers of V1. One pathway seems to deal primarily with color and this is called the P-B pathway. Neurons in the second pathway are sensitive to features such as the orientation of the stimulus and seem to mediate high acuity perception. This pathway is called the P-I pathway. The M pathway is dominated by a single source, but the P-I and P-B streams receive inputs from a number of different sources. At the top of the cortical hierarchy, the M pathway leads primarily to the posterior parietal cortex, which processes spatial and motion information. The P-I and P-B pathways project to the inferior temporal cortex, which mediates pattern and object recognition. Abbreviations: RGC, retinal ganglion cell; LGN, lateral geniculate nucleus; V1 to V5, visual areas 1 to 5; PPC, posterior parietal cortex; IT, inferior temporal cortex. (Redrawn from Van Essen DC, Deyoe, EA. In The Cognitive Neurosciences. Gazzanida MSEC(ed). London: MIT press 1995; 384-400.).

direction of movement. Some of these neurons are binocular (require stimulation from both eyes) and show sensitivity to retinal disparity (the difference in the relative position of the stimuli in the visual field of two eyes).[19] But these cells do not unambiguously signal stereoscopic depth, although combining the outputs of many V1 neurons could potentially do so.[20] The disparity signals may also be important for the rapid involuntary control of vergence eye movements (eye movements that bring the images on the two foveae into register).[21]

The P pathway splits to produce two new pathways in the upper layers of V1. One pathway seems to deal primarily with color and this is called the P-B pathway. Neurons in the second pathway are sensitive to features such as the orientation of the stimulus and seem to mediate high acuity perception. This pathway is called the P-I pathway. This separation been demonstrated anatomically by staining for the mitochondrial enzyme cytochrome oxidase. In both monkeys and humans, this stain shows an arrangement of dark-staining columns which extend through layers 2 and 3, and more faintly through layers 5 and 6.[22] These columns, called blobs, are 0.2 mm in diameter and spaced at roughly 0.5 mm intervals. The areas surrounding the blobs are called the interblob region. In the interblob region, most cells respond to stimuli of a particular orientation, such as lines or bars, and have small receptive fields. Most of these cells do not show color coded responses. They show no color opponency and respond well to achromatic luminance contrast borders. This suggests that the color-coded P cell input is pooled in such a way that color contrast can be used to identify borders, but that the information about the colors forming the border is lost. The neurons in the interblob region are part of the P-I pathway.

The blob cells are not orientation selective, but are either color or brightness selective. These cells are part of the P-B pathway. The P-B pathway thus seems to carry information complementary to the information carried by the P-I pathway. The color-opponent blob cells receive input from the color-opponent P cells in the LGN, although they differ in that their receptive field centers are larger and their color coding is double opponent (they give opposite responses to different parts of the spectrum in the different parts of their receptive field. For example, the center might give an on-response to green and an off-response to red, and the opposite set of responses in the surround).

The blob and interblob systems thus seem to work in different but complementary ways. The blob cells are color-coded, excited by colors in one region of the spectrum and inhibited by others, and not selective for stimulus orientation. Interblob cells are selective for stimulus orientation, but mostly are not color selective, responding to a line or edge of the correct orientation regardless of its color.

The main target of V1 is V2. Staining for cytochrome oxidase in this area does not reveal a pattern of blobs or interblobs, but instead a pattern of stripes running perpendicular to the border between V1 and V2, and which extends over the entire 8 to 10 mm width of V2. There seem to be three types of stripe. There are two darkly staining stripes, one thick and one thin, separated by more lightly staining interstripes. (sometimes called pale stripes). The neurons in layer 4B of V1 (part of the M pathway) project to the thick stripes. Neurons in the thick stripes show similar response properties to the neurons in layer 4B. They are orientation and movement selective, and many show sensitivity to retinal disparity. The neurons in the blobs project to the thin stripes, and neurons in the thin stripes are not orientation selective, but more than half are color sensitive (mostly double-opponent). The interblobs project to the interstripes, and neurons in this region are orientation selective but not direction selective, nor do they show color selectivity. The organization of V1 is retinotopic; that is, the visual field of the retina is mapped onto the surface of the cortex of V1. In V2 there seem to be three separate visual maps.[22] Within the thick stripes there is an visual orientation map, within the thin stripes there is a color map, and within interstripes a disparity map. Adjacent stripes are responsive to the same region of visual field. So there are three, interleaved visual maps in V2, each representing a different aspect of the visual stimulus.

Thus, the M pathway projects from layer 4B of V1 to the thick stripes of V2. The P-B pathway projects from the blobs of V1 to the thin stripes of V2, and the P-I pathway projects from the interblob region to the interstripes.

Extrastriate Processing

Both subdivisions of the P pathway, the thick stripes (color) and interstripes (form) project to visual area 4 (V4). V4 and the other visual areas upstream of V2 all stain relatively homogeneously for the cytochrome oxidase enzyme, and no alternative marker is yet known. However, the continued separation of the two subdivisions of the P pathway in V4 can be inferred from patterns of connectivity. It is possible to trace connections by using substances, such as peroxidase enzymes or dyes, which will be absorbed by the neurons and transported up or down their axons. This method reveals the neuronal connections of the specific piece of cortex in which the tracer is deposited. When tracer is deposited in different parts of V4, the back-filled neurons in V2 tend to occur in one of two distinct patterns; either they are largely restricted to the thin stripes or they are largely restricted to the interstripes.[23-24] V4 projects primarily to posterior inferior temporal cortex (PIT). Retrograde tracing from this area shows a patchy, modular organization of the cells in V4,[25] a finding supported by a combined electrophysiological and optical imaging study.[26] This suggests that the anatomical separation of the two subdivisions of the P pathway continues in V4.

V4 has received a great deal of attention as the putative site for the development of color constancy. This region has been functionally explored primarily in two ways in the macaque: lesions and electrophysiology. The results of lesion studies suggest a dissociation between hue discrimination and color constancy. While monkeys with V4 lesions are not significantly impaired on hue discrimination tasks,[27-29] they are impaired on color constancy tasks.[27-28] Yet the most striking deficit caused by V4 ablation in monkeys is in pattern discrimination. These facts emphasize that V4 cannot be solely devoted to analyzing color, and also suggest that V4 cannot be the homologue of the putative color center damaged in human cerebral achromatopsia, since such achromatopsics are unimpaired in pattern discrimination but are severely impaired on hue discrimination tasks.

However, lesion experiments contain a number of intrinsic drawbacks, of which the most difficult to ameliorate is that the removal of an area from the complex network of forward, lateral and back connections that make up the visual pathways does alter the way the whole visual system works. To understand how the visual system functions normally, the best technique presently available is microelectrode recording. This technique has shown that neurons in V4 display a number of characteristics which suggest a more elaborate representation of chromatic information than in preceding areas, and would be consistent with the development of color constancy. For example, V4 neurons have comparatively large receptive fields, sometimes extending into the ipsilateral visual field, suitable for carrying out the summation of inputs from a large region of retinotopic space.[30-32] Moreover, consistent with the computation of color contrasts, some V4 cells show spectral selectivity for stimuli outside the classical receptive field.[31-33] These properties are necessary for the computation of color constancy, according to all theories that require long-range spatial interactions or estimates of the illuminant based on samples taken across the image.[34-36]

In a now classic series of studies, Zeki,[30,32,37] employed a color Mondrian stimulus to search for neuronal correlates of color constancy in V1 and V4 of anesthetized macaques. He reported that whereas neurons in V1 responded to the simple wavelength composition of stimuli, neurons in V4 appeared to signal the color perceived by a human observer, irrespective of the illuminant. This suggested the emergence in V4 neurons of new chromatic properties which are not seen in cells of previous regions, and which support the perception of color constancy.

There may, though, be alternative explanations of the results. The effects described in V4 depend on stimulating regions surrounding the classical receptive field,[33] but Maunsell and Newsome[38] have suggested that the surrounds of the receptive fields in V1 were not stimulated in an equivalent manner. Mondrian stimuli of fixed size were used to test neurons in both V1 and V4. Because receptive fields in V1 cover only 1/30 of those in V4, a single patch of the Mondrian may have covered a critical region of the surround, as well as the classical receptive field of the V1 neurons. The observed differences between V1 and V4 may therefore simply reflect differences in the spatial extent of the neuron's classical receptive fields and their surrounds, rather than the appearance of a new property in V4.[38]

The M pathway projects to visual areas 3 (V3) and 5 (V5), both directly from layer 4B of V1 and through the thick stripes of V2. Most cells in V3 are motion and orientation selective and are believed to be concerned with processing dynamic form.[39-40] V5 (also known as the middle temporal visual area or MT) is believed to process information on motion and stereoscopic depth. In monkeys, lesions of V5 cause deficits in pursuit eye movements and in discriminating the direction of motion. The M pathway then projects to the parietal cortex.. The parietal cortex seems to be important for the integration of movement and depth into a representation of space. Damage to this region in humans causes a condition called Balint's Syndrome. This has three main symptoms. Firstly, there is a difficulty in reaching for objects under visual guidance (optic ataxia). Secondly, subjects display a deficit in visual scanning (ocular apraxia). A person may perceive an object normally, but will be unable to maintain fixation. His or her eyes will wander. He or she will not be able to make a systematic scan of a scene and will not be able to perceive the location of the objects seen. Finally, the subject is only able to see one object at a time in a crowded scene (simultagnosia).

V5 seems to be divided into two subdivisions which analyze different aspects of motion, which seem to be related to two broad areas of function. These are the motion of an object through the environment and the motion effects caused by our own movement through the environment.[16] These subdivisions project to separate visual areas within the parietal lobe; medial superior temporal (MST)l and MSTd. Neurons in MSTl seem to be responsive to the motion of an object through the environment, whereas neurons in MSTd seems sensitive to motion caused by movement of our eyes or of ourselves. Neurons in this latter area are responsive to changes in certain parameters of a stimulus, such as an increase or decrease in its size (such as might be produced by moving toward or away from it), its rotation (such as might be produced when tilting our heads) and shear (such as might be produced when moving past objects at different distances).[41-43] In addition, some cells are sensitive to mixtures of these stimulus parameters, such as spiral motion patterns which have components of both rotation and expansion.[44] Thus, cells in MSTd seem ideally suited to encode the visual changes which occur when we move, and allow us to interact with our environment.

The separation of the P and M pathways should not be overemphasized. There is some communication between the two pathways. If the M layers of the LGN are inactivated by cooling, the visual responses of neurons are reduced both in V4 and in the blobs and interblobs of V1.[45-46] This suggests that there is an input from the M pathway into the P pathway. Moreover, there is a third type of cell in the LGN. These small neurons are called K or W cells, and are found within the gaps between the M and P layers. They are believed to receive input from Pγ cells, one of the cell types which make up the remaining 10% of retinal ganglion cells. The W neurons also project to the blobs of V1.[47] Thus, the so called P pathway actually receives input from the P, M and W neurons of the LGN. The M pathway does seem to be more segregated. For example, inactivation of the P layers of the LGN has a negligible effect on the responses of neurons in V5.[48]

The Functional Organization of the Visual System

Visual information seems to be processed in two broad systems in both Old and New World Monkeys,[9,49] and possibly also in the cat.[50] However, there is some disagreement about the exact transforms undertaken in these systems. In a modification of a model originally proposed for rodent visual processing,[51] Mishkin proposed that visual processing can be divided into a "what" system (also called the ventral system), which is concerned with the identification of an object, and the "where" system (also called the dorsal system), which is concerned with the relative spatial position of an object[52] (Fig. 1.4). Damage to the "what" system, through which V1 projects to the temporal lobes, does not impair performance of visuospatial tasks, but does impair performance of object discrimination tasks.[53-54] Damage to the "where" system, through which the V1 projects to the parietal lobes, produces impairments on visuospatial tasks, but does not impair the performance of object discrimination tasks.[55-57] The M and P pathways have been thought to correspond roughly to these two systems.[58] The P pathway, with information about color and shape, would seem to be ideal for the "what" system, which develops a representation of an object. Similarly, the M pathway, with information about motion, stereopsis and form derived from motion, would seem to be the obvious candidate for the "where" system, which develops a representation of spatial relationships in the visual field. However, there is considerable communication between the two systems at all levels.[59]

An alternative approach has been proposed which takes more account of the output requirements of the system.[60-61] This approach is called "what" versus "how", and has been supported both by primate electrophysiology and clinical evidence. Shape-selective neurons have been reported in the intraparietal sulcus.[62-64] Many cells in the lateral bank of the sulcus selectively respond to images of a switch, a lever, a button or a knob in a groove, that the monkeys had been trained to manipulate. The hand movements and the shape of the grasp required to manipulate these objects differed (pulling the lever, pushing the button, grasping and pulling the knob, etc.), and the monkey moved its arm and shaped its hand to specifically manipulate each object before the monkey reached out. The responses of neurons in the lateral sulcus started when the monkey saw the object, and the response was reduced if the monkey performed the task in a dark room. This suggests that the neural responses might

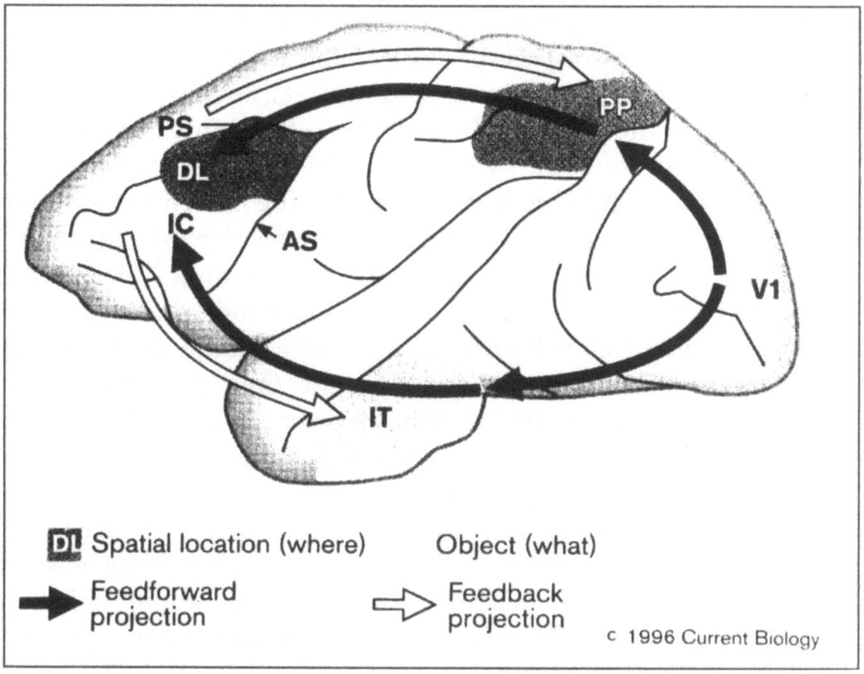

Fig. 1.4. The location of the "what" and 'where" pathways in the primate brain, and the proposed feedback projections from prefrontal cortex during working memory tasks. PS, principal sulcus; AS, arcuate sulcus; PP, posterior parietal cortex; IT, inferior temporal cortex; DL, dorsolateral frontal cortex; IC, inferior convexity of the frontal cortex; V1, primary visual cortex. (Reproduced with permission from Tovée MJ, Cohen-Tovée EM. Current Biology 1996; 6: 13-15. Current Biology).

be partially stimulated by visual inputs. Additionally, there are cells in a more posterior part of the sulcus which respond to more primitive features of stimuli, including the 3D orientation of a pole or a 3D tilt of a plane.[65-66] The responses were reduced when the degree of binocular disparity was reduced, suggesting that the binocular disparity represents an important cue for these responses. These results suggest that in the parietal cortex there is a synthesis of action-oriented visual information with other sensory cues to produce a basis for visually-directed movement.[61,66]

Visually guided grasping was studied in a patient with Balint's Syndrome (as mentioned above, in this syndrome bilateral parietal damage causes profound disorders of spatial attention, gaze and visually guided reaching). While this patient had no difficulty in recognizing line drawings of common objects, her ability to pick up objects remained impaired.[67] Not only did she fail to show normal scaling of the grasping movement, she also made a large number of adjustments in her grasp as she closed in on an object. These adjustments are rarely observed in normal subjects. Such studies suggest that damage to the parietal lobe can impair the ability of patients to use information about the size, shape and orientation of an object to control the hand and fingers during a grasping movement. Another patient developed profound visual-form agnosia (an inability to recognize objects) following carbon monoxide poisoning.[68] Despite her profound inability to recognize the shape, size and orientation of an object, the patient showed strikingly accurate guidance of hand and finger movements directed at the very same objects. So despite impaired conscious visual discrimination of the objects, visual information, computed unconsciously, was made available to the action system to direct grasping actions.

The two streams project to different prefrontal cortical areas.[69] The "what" system projects to the cortex of the inferior convexity (IC) ventrolateral to the principal sulcus and the "where" or "how" system projects to the dorsolateral prefrontal region (DL). The prefrontal cortex is an important region for working memory. Goldman-Rakic and her colleagues trained monkeys to carry out visual tasks, while at the same time they recorded from single neurons in either IC or DL. In the first task, the monkeys were trained to stare at a spot on a video screen while an image flashed at one of several locations on the screen and then disappeared. A few seconds later, a cue on the screen signaled the monkeys to move their gaze to where the image had been, indicating that they had remembered the location of the image. In the second task, the location of the image remained constant, but the image itself changed. The monkeys were trained to wait until the image disappeared, and then, after a delay, move their eyes to the right if they saw one image and to the left if they saw another, indicating they remember information about an object's features. In the first task, neurons in DL became active during the delay period, while neurons in IC did not alter their activity. However, during the second task the pattern of activity is reversed; the neurons in IC are active during the delay and the neurons in DL remain quiet. These results suggest that IC mediates working memory for objects and Dl mediates spatial working memory. However, an alternative view is that the prefrontal cortex may be a site where spatial and form information are integrated, and that the apparent segregation is a product of the way response properties of the neurons were measured.[70]

Visual Working Memory

In the highest association areas, the activity of the neurons plays an important role in working memory. For example, in the primate inferior temporal (IT) cortex there are neurons responsive to complex biological stimuli such as faces. A particular stimulus is encoded by the unique pattern of activity across a population of these neurons.[71-72] Such neurons are not only important for the recognition of a stimulus, but also seem to play a role in the maintenance of visual working memory. The effect of the prior presentation of a visual stimulus on the responses of these neurons can be recorded in either of two ways: by suppression of neural response or by enhancement of neural response. Repeated presentation of a particular stimulus reduces the responses of IT neurons to it, but not to other stimuli. This selective suppression of neural responses to familiar stimuli may function as a way of making new or unexpected stimuli stand out.[73] This selective suppression can be found in IT cortex in monkeys passively viewing stimuli and even in anesthetized animals, suggesting that it is an automatic process which is independent of cognitive factors.

Enhancement of neural activity has been reported to occur when a monkey is actively carrying out a working memory task, such as delayed matching to sample (DMS). In the basic form of this task, a sample stimulus is presented, followed after a delay, by a test stimulus. The monkey has to indicate whether the test stimulus matches or differs from the sample stimulus. Some neurons in monkey IT maintain a high firing rate during the delay between stimuli (which can be as long as 10-15 sec), as though they are actively maintaining a memory of the sample stimulus for comparison with the test stimulus.[74] However, if a new stimulus is presented during the delay between the sample and test, the maintained neural activity is abolished.[75] This neural activity seems to represent a form of visual rehearsal which can be easily disrupted, but it may still be an aid to short term memory formation.[73]

In another form of the DMS task, a sample stimulus was presented followed by a sequence of test stimuli and the monkey had to indicate which of these matched the sample.[76] Under these conditions, a proportion of IT neurons gave an enhanced response to the test stimulus that matched the sample stimulus. The basis of this enhanced response lies in signals coming in a top-down direction from the ventral prefrontal cortex, an area which has been implicated in short term visual memory.[69] Like IT neurons, some neurons in IC show a maintained firing rate during the delay interval.[73,77] This maintained firing is temporarily interrupted by additional stimuli shown during the delay interval, but the activity rapidly recovers. This maintained information about the sample stimulus may be fed back from the prefrontal cortex to the IT neurons so that they give an enhanced response to the correct test stimulus.[73] Consistent with this hypothesis is the finding that performance of the task can be reversibly disrupted by temporarily cooling either the prefrontal or inferotemporal cortex.[78-79] Moreover, cooling one of these cortices affects cell activity in the other,[80] which argues for an important role for recurrent connections between these two areas (Fig. 1.4).

A similar feedback system seems to function in the dorsal stream. Neurons in posterior parietal cortex (PPC) and DL are sensitive to the spatial relationships in the environment. A recent study has shown that there is coactivation of these areas in spatial memory tasks,[81] and the reversible inactivation of either area through cooling leads to deficits in such tasks.[82] Neurons in both areas show a maintained response during the delay interval like those in IT and IC, and the maintained activity in PPC can be disrupted by cooling of DL.[83] This suggests that feedback from prefrontal areas is important for the maintenance of the neuronal activity in the higher visual association areas associated with visual working memory.

Fine-Tuning Memory

It is well known that there is extrathalamic modulation of the cortical visual system at all levels and this includes the prefrontal cortex and the higher association areas.[84] Recent studies have concentrated on the dopamine innervation of the prefrontal area, and it has been shown that changes in dopamine levels are associated with working memory deficits in monkeys.[85] These studies have an immediate relevance to clinical treatment, as changes in the dopamine innervation of the prefrontal cortex have been implicated in working memory deficits in both Parkinson's disease and schizophrenia. The prefrontal cortex is a major target of the brain stem dopamine afferents which synapse onto the spines of pyramidal neurons.[86] The same spines often also have excitatory synapses from the sensory inputs arriving at the prefrontal cortex, and this arrangement has the potential to allow direct dopamine modulation of local spinal responses to excitatory input.[87] A particular subtype of dopamine receptor (D_1) is concentrated in the prefrontal cortex, primarily on the spines of pyramidal cells,[88] and iontophoresis of a D_1 antagonist enhances the activity of neurons in DL during the intertrial of a spatial memory task. These DL neurons seem to display spatially tuned "memory fields".[89] The neurons would respond maximally during the delay period to targets which had appeared in one or a few adjacent locations, but would not respond above baseline for any other location. Different neurons seem to code for different spatial locations, and so it is possible that a precise spatial location could be encoded across a population of neurons. The D_1 antagonist causes an enhancement of the delay activity for stimuli in a cell's memory field, but not for any target locations outside this memory field. This effect is dose-dependent. Higher levels of D_1 antagonists inhibited cell firing at all stages of the spatial memory task; it did not matter whether the target stimulus was shown in the memory field or outside it. These results suggest that intensive D_1 receptor blockade may render prefrontal cells unresponsive to their normal functional inputs, and Williams and Goldman-Rakic[89] suggest that this may be through indirect mechanisms involving inhibitory local circuits. This possibly explains the reports that deficits in working memory are produced by injection of D_1 antagonists,[90] and the reports of inhibition of delay period activity by nonselective dopamine antagonists.[91]

References

1. Rakic P. Specification of cerebral cortical areas. Science 1988; 241:170-176.
2. Rakic P. A small step for the cell a giant leap for mankind: A hypothesis of neocortical expansion during evolution. Trends in Neurosciences 1995; 9:383-38.
3. Griffin L. The intrinsic geometry of cerebral cortex. J Theoretical Biol 1994; 166:261-273.
4. Scannell JW. Determining cortical landscapes. Nature 1997; 386:452.
5. Van Essen DC. A tension-based theory of morphogenesis and compact wiring in the central nervous system. Nature 1997; 385:313-318.
6. Cowey A. Cortical maps and visual perception. Quart J Exp Psychol 1979; 31:1-17.
7. Mitchison G. Neuronal branching patterns and the economy of cortical wiring. Proc R Soc Lond B 1991; 245:151-158.
8. Cherniak C. Local optimization of neuron arbors. Biol Cyber 1991; 66:503-510.
9. Young MP. Objective analysis of the topological organization of the primate cortical visual system. Nature 1992; 358:152-154.
10. Cherniak C. Neural component placement. Trends in Neurosciences 1995; 18:522-527.

11. Scannell JW, Blakemore C, Young MP. Analysis of connectivity in the cat cerebral cortex. J Neurosci 1995; 15:1463-1483.
12. Cherniak C. Component palcement optimisation in the brain-reply.Trends in Neurosciences 1996; 19:414-415.
13. Young MP, Scannell JW. Component placement optimisation in the brain. Trends in Neurosciences 1996; 19:413-414.
14. Scannell JW, Young MP. The connectional organization of neural systems in the cat cerebral-cortex. Current Biology 1993; 3:191-200.
15. Kaas JH. Why does the brain have so many visual areas? J Cog Neurosci 1989; 1:121-135.
16. Van Essen DC, Gallant JL. Neural mechanisms of form and motion processing in the primate visual system. Neuron 1994; 13:1-10.
17. Kaas JH. Human visual cortex: Progress and puzzles. Current Biology 1995; 5:1126-1128.
18. Sereno MI, Dale AM, Reppas JB, Kwong KK, Belliveau JW, Brady TJ, Rosen BR, Tootell RBH. Borders of multiple visual areas in humans revealed by functional magnetic resonance imaging. Science 1995; 268:889-893.
19. Poggio GF, Fischer B. Binocular interaction and depth sensitivity in striate cortex and prestriate cortex of behaving rhesus monkey. J Neurophysiol 1977; 40:1392-1405.
20. Cumming BG, Parker AJ. Responses of primary visual cortical neurons to binocular disparity without depth perception. Nature 1997; 389:280-283.
21. Masson GS, Busettini C, Miles FA. Vergence eye movements in response to binocular disparity without depth perception. Nature 1997; 389:283-286.
22. Roe AW, Ts'o D. Visual topography in primate V2: Multiple representation across functional stripes. J Neurosci 1995; 15:3689-3715.
23. Shipp S, Zeki S. Segregation of pathways leading from area V2 to areas V4 and V5 of the macaque monkey visual cortex. Nature 1985; 315:322-325.
24. Zeki S, Shipp S. The functional logic of cortical connections. Nature 1988; 335:311-317.
25. Felleman DJ, Xiao YP, McClendon E. Modular organization of occipito-temporal pathways: Cortical connections between visual area 4 and visual area 2 and posterior inferotemporal ventral area in macaque monkeys J Neurosci 17:3185-3200.
26. Ghose GM, Ts'o DY. Form processing modules in primate area V4. J Neurophysiol 1997; 77:2191-2196.
27. Wild HM, Butler SR, Carden D, Kulikowski JJ. Primate cortical area V4 important for color constancy but not wavelength discrimination. Nature 1985; 313:133-135.
28. Walsh V, Carden D, Butler SR, Kulikowski JJ. The effects of V4 lesions on the visual abilities of macaques: Hue discrimination and color constancy. Behav Brain Res 1993; 53:51-62.
29. Heywood CA, Gadotti A, Cowey A. Cortical area 4 and its role in the perception of color. J Neurosci 1992; 12:4056-4065.

30. Zeki S. The representation of colors in the cerebral cortex. Nature 1980; 284:412-418.
31. Zeki S. The visual image in mind and brain. Scientific American 1992;.267:69-76.
32. Schein SJ, Desimone R. Spectral properties of V4 neurons in the macaque. J Neurosci 1990; 10:3369-3389.
33. Zeki S. Color coding in the cerebral cortex: The responses of wavelength-selective and color-coded cells in monkey visual cortex to changes in wavelength composition. Neuroscience 1983; 9:741-765.
34. Land EH. Recent advances in retinex theory and some implications for cortical computations. Proc Natl Acad of Sci USA 1983; 80:5163-5169.
35. Hurlbert A. Color-vision—putting it in context. Current Biology 1996; 11:1381-1384.
36. Maloney LT, Wandell BA. Color constancy—a method for recovering surface spectral reflectance. J Opt Soc Am A 1986; 3:29-33.
37. Zeki S. Color coding in the cerebral cortex: The reaction of cells in the monkey visual cortex to wavelengths and colors. Neuroscience 1983; 9:741-765.
38. Maunsell JHR, Newsome WT. Visual processing in monkey extrastriate cortex. Ann Rev Neurosci 1987; 10:363-401.
39. Zeki S. A Vision of the Brain. Blackwell Scientific Publications. 1993.
40. Gegenfurtner KR, Kiper DC, Levitt JB. Functional properties of neurons in macaque area V3. J Neurophysiol 1997; 77:1906-1923.
41. Saito H-A, Yuki M, Tanaka K, Hikosaka K, Fukada Y, Iwai E. Integration of direction signals of image motion in the superior temporal sulcus of the macaque. J Neurosci 1986; 6:145-157.
42. Duffy CJ, Wurtz RH. Sensitivity of MST neurons to optic flow stimuli I. A continuum of response selectivity to large field stimuli. J Neurophysiol 1991; 65:1329-1345.
43. Orban GA, Lagae L, Verri A, Raiguel S, Xiao D, Maes H, Torre V. First-order analysis of optical flow in monkey brain. Proc Natl Acad Sci USA 1992; 89:2595-2599.
44. Graziano MSA, Andersen RA, Snowden RJ. Tuning of MST neurons to spiral motions. J Neurosci 1994; 14:54-67.
45. Nealey TA, Maunsell JHR. Magnocellular and parvocellular contributions to the responses of neurons in the macaque striate cortex. J Neurosci 1994; 14:2080-2088.
46. Ferrera VP, Neally TA, Maunsell JHR. Responses in macaque visual area V4 following the inactivation of the parvocellular and magnocellular LGN pathways. J Neurosci 1994; 14:2080-2088.
47. Hendry SHC, Yoshioka T. A neurochemically distinct third channel in the macaque dorsal lateral geniculate nucleus. Science 1994; 264:575-577.
48. Maunsell JHR, Nealey TA, DePriest DD. Magnocellular and parvocellular contributions to responses in the middle temporal visual area (MT) of the macaque monkey. J Neurosci 1990; 10:3323-3334.

49. Weller RE. Two cortical visual systems in Old World and New World primates. In: Hicks TP, Benedek G eds. Progress in Brain Research Vol 75, Amsterdam: Elsevier 1988; 293-306.
50. Lomber SG, Payne BR,Cornwall P, Long KD. Perceptual and cognitive visual functions of parietal and temporal cortices in the cat. Cereb Cortex 1996; 6:673-695.
51. Schneider GE. Two visual systems: Brain mechanisms for localization and discrimination are dissociated by tectal and cortical lesions. Science 1969; 163:895-902.
52. Mishkin M, Ungerleider LG, Macko KA. Object vision and spatial vision: Two cortical pathways. Trends in Neurosciences 1983; 6:414-417.
53. Pohl W. Dissociation of spatial discrimination deficits following frontal and parietal lesions in monkeys. J Comparative Physiol 1973; 82:227-239.
54. Mishkin M, Lewis ME, Ungerleider LG. The equivalence of parieto-preoccipital subareas for visuospatial ability in monkeys. Behav Brain Res 1982; 6:57-77.
55. Ungerleider LG, Mishkin M. Two coritcal visual systems. In: Ingle DJ, Goodale MA, Mansfield RJ, eds. The Analysis of Visual Behaviour 1982; 549-586. Cambridge: MIT Press.
56. Gaffan D, Harrison S, Gaffan EA. Visual identification following interotemporal ablation in the monkey. Quart J Exp Psychol 1986; 30B:5-30.
57. Weiskrantz L, Saunders RC. Impairments of visual object transforms in monkeys. Brain 1984; 107:1033-1072.
58. Livingstone M, Hubel D. Segregation of form color movement and depth: Anatomy physiology and perception. Science 1998; 240: 740-749.
59. Harries MH, Perrett D1. Visual processing of faces in the temporal cortex: Physiological evidence for a modular organization and possible anatomical correlates. J Cog Neurosci 1991; 3:9-24.
60. Goodale MA, Milner AD. Separate visual pathways for perception and action. Trends in Neuroscience 1992; 15:20-25.
61. Sakata H, Taira M. Parietal control of hand action. Curr Opin Neurobiol 1994; 4:847-856.
62. Taira M, Mine S, Georgopoulos AP, Murata A, Sakata H. Parietal cortex neurons of the monkey related to the visual guidance of hand movement. Exp Brain Res 1990; 83:29-36.
63. Sakata H, Kusonoki M. Organization of space perception: Neural representation of three-dimensional space in the parietal cortex. Curr Opinion Neurobiol 1992; 2:170-174.
64. Sakata H, Shibutani H, Ito Y, Tsurugai K, Mine S, Kusunoki M. Functional-properties of rotation-sensitive neurons in the posterior parietal association cortex of the monkey. Exp Brain Res 1994; 101:183-202.
65. Sakata H, Taira M, Murata A, Mine S. Neural mechanisms of visual guidance of hand action in the parietal cortex of the monkey. Cereb Cortex 1995; 5:429-438.

66. Sakata H. Coding of 3-D features used in manipulation by parietal neurons In: Caminiti R, Hoffman K-P, Lacquaniti F, Altman J eds. Vision and Movement Mechanisms in the Cerebral Cortex. Strasbourg: HFSP 1996; 55-63.
67. Jakobsen LS, Archibald YM, Carey DP, Goodale MA. A kinematic analysis of reaching and grasping movements in a patient recovering from optic ataxia. Neuropsychologia 1991; 29:803-809.
68. Goodale MA, Milner AD, Jacobsen LS, Carey DP. A neurological dissociation of between perceiving objects and grasping them. Nature 1991; 349 154-156.
69. Wilson FAW O', Scalaidhe SP, Goldman-Rakic PS. Dissociation of object and spatial processing domains in primate prefrontal cortex. Science 1993; 260:1955-1958.
70. Rao SC, Rainer G, Miller EK. Integration of what and where in the primate prefrontal cortex. Science 1997; 276:821-824.
71. Tovée MJ, Rolls ET. Information encoded in short firing rate epochs by single neurons in the primate temporal visual cortex. Visual Cognition 1995; 2:35-58.
72. Abbott LF, Rolls ET, Tovée MJ. Representational capacity of face coding in monkeys. Cereb Cortex 1996; 6:498-505.
73. Desimone R, Miller EK, Chelazzi L, Lueschow A. Multiple visual systems in the visual cortex. The Cognitive Neurosciences ed Gazzanida 1995; 475-486.
74. Miyashita Y, Chang HS. Neuronal correlate of pictorial short-term memory in the primate temporal cortex. Nature 1988; 331 68-70.
75. Baylis GC, Rolls ET. Responses of neurons in short-term and serial recognition memory tasks. Exp Brain Res 1987; 65:614-622.
76. Miller EK, Desimone R. Parallel neuronal mechanisms for short-term memory. Science 1994; 263:520-522.
77. Miller EK, Erickson CA, Desimone R. Comparison of prefrontal (PF) and inferior temporal (IT) neurons during performance of a memory task. Soc Neurosci Abstr 1995; 21:661-663.
78. Bauer RH, Fuster JM. Delayed matching and delayed-response deficit from cooling dorsolateral prefrontal cortex in monkeys. J Comp Physiol Psychol 1976; 90:293-302.
79. Fuster JM, Bauer RH, Jervey JP. Effects of cooling inferotemporal cortex on performance of visual memory tasks. Exp Neurol 1981; 71:398-409.
80. Fuster JM, Bauer RH, Jervey JP. Functional interations between inferotemporal and prefrontal cortex in a cognitive task. Brain Research 1985; 330:299-307.
81. Friedman HR, Goldman-Rakic PS. Coactivation of prefrontal cortex and inferior parietal cortex in working memory tasks revealed by 2DG functional mapping in the rhesus monkey. J Neurosci 1994; 14:2775-2788.
82. Quintana J, Fuster JM. Spatial and temporal factors in role of prefrontal and paratiel cortex in visuomotor integration. Cereb Cortex 1993; 3:122-132.

83. Goldman-Rakic PS, Chafee M. Feedback processing in prefronto-parietal circuits during memory-guided saccades. Soc Neurosci Abstr 1994; 20:808.
84. Foote SL, Morrison JH. Extrathalamic modulation of cortical function. Ann Rev Neurosci 1986; 10:67-95.
85. Robbins TW, Roberts AC, Owen AM, Sahakian BJ, Everitt BJ, Wilkinson L, Muir J, DeSalvia M, Tovée MJ. Monoaminergic-dependent cognitive functions of the prefrontal cortex in monkey and man. In: Motor and Cognitive Functions of the Prefrontal Cortex, Thiery AM et al, eds. Berlin: Springer-Verlag 1994; 93-111.
86. Williams GV, Goldman-Rakic PS. Characterization of the dopamine innervation of the primate frontal cortex using a dopamine specific antibody. Cereb Cortex 1993; 3:199-222.
87. Goldman-Rakic PS. Cellular basis of working memory. Neuron 1995; 14:477-485.
88. Smiley JF, Levy AI, Ciliax BJ, Goldman-Rakic PS. Dopamine receptor immunoreactivity in in human and monkey cerebral cortex: Predominant and extrasynaptic localization in dendritic spines. Proc Natl Acad Sci USA 91:5720-5724.
89. Williams GV, Goldman-Rakic PS. Modulation of memory fields by dopamine D1 receptors in prefrontal cortex. Nature 1995; 376:572-575.
90. Arnsten AFT, Cai JX, Murphy BL, Goldman-Rakic PS. Dopamine D1 receptor mechanisms in the cognitive performance of young-adult and aged monkeys. Psychopharmacology 1994; 116:143-151.
91. Sawaguchi T, Matsumura M, Kubota K. Effects of dopamine antagonists on neural activity related to a delayed-response task in monkey prefrontal cortex. J Neurophysiol 1990; 63:1401-1412.

CHAPTER 2

The Representation of Complex Stimuli

The Development of Response Invariance

In the primary stages of the visual system, such as VI, objects are coded in terms of retinotopic coordinates, and lesions of VI cause defects in retinal space which move with eye movements, maintaining a constant retinal location. Several stages later in the visual system, at the inferior temporal cortex, the receptive fields are relatively independent of retinal location, and neurons can be activated by a specific stimulus, such as a face, over a wide range of retinal locations. Deficits that result from lesions of the inferior temporal cortex are based on the coordinate system properties of the object, independent of retinal location. Thus, at some point in the visual system, the pattern of excitation that reaches the eye must be transposed from a retinotopic coordinate system to a coordinate system centered on the object itself.[1,2]

At the same time that coordinates become object centered, the system becomes independent of the precise metric regarding the object itself within its own coordinate system, that is to say that the system remains responsive to an object despite changes in its size, orientation, texture, and completeness.[2] Single-cell recording studies in the macaque suggest that these transformations occur in the inferior temporal cortex (IT).[3-4] So, for example, neurons in IT should respond to a stimulus no matter where it is in the visual field (translation invariance). A demonstration of translation invariance comes from a study by Tovée, Rolls and Azzopardi.[5] The responses of 44 single neurons in the inferior temporal cortex and the cortex in the banks of the anterior part of the superior temporal sulcus of three awake, behaving macaques were recorded during a visual fixation task.[5] Stimulus images subtending 17° were presented in the center of a VDU screen, and fixation was either at the center of the screen, or at one of four positions around the edge of the test stimulus. The responses were compared for fixation at each position for both effective and noneffective stimuli for each cell.

An example of the effects of different fixation positions on the responses of a neuron to two different stimuli (one effective and one noneffective) is shown in Figure 2.1. The face subtended 17° at the retina. The means and standard errors of the means of the neuronal responses at the different fixation positions are shown. The neuronal responses were as large when the monkey was fixating at the edge of an effective stimulus for the cell as when he was fixating at the center of the same stimulus (Fig. 2.1A). The stimulus selectivity of the cell was maintained across this area of visual field. A less effective stimulus for the cell elicited only a small response when the monkey was fixating at the center of that stimulus and at its edges (Fig. 2.1B).

The firing rates of most neurons to an effective image did not significantly alter when visual fixation was as far eccentric as the edge of the stimulus, and they showed only a small reduction when the fixation point was up to 4° from the edge of the stimulus. Moreover, stimulus selectivity across faces was maintained throughout this region of the visual field. Thus, the neurophysiological data show that the responses of the cells were relatively independent of fixation position. Analysis of the neuronal spike trains using information theory was next used to quantify the extent to which information was made explicit in the cells' responses about which stimulus was shown, rather than information about where the stimulus was on the retina.

Using principal component analysis and information theory, it was possible to measure for each neuron the information about stimulus type and fixation position contained in the firing rate measured over a poststimulus period of 0-400 msec (Fig. 2.2).[6-7] This method is discussed in detail in chapter 3. The information theoretic analysis also assessed how much information about image type and fixation position was carried by the temporal properties of the spike train (Fig. 2.3). This population of cells had considerable information about which of four stimuli were shown (on average 0.203 bits), and only about one-sixth as much information about where the fixation point was on the stimuli (on average 0.035 bits). This is made more significant when one considers that, as there were more fixation positions (5) than stimuli (4), the maximum information a cell could have conveyed about the 4 stimuli was 2 bits, and about 5

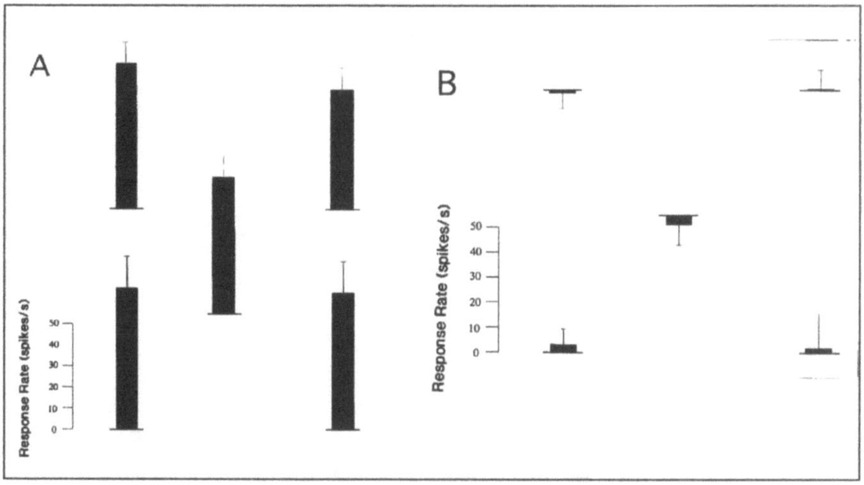

Fig. 2.1. Neuronal responses for a cell with different fixation positions when a full-size effective stimulus (A) and a full-sized noneffective stimulus (B) were presented. The means and standard errors of the neuronal responses (firing rate minus spontaneous rate) are shown. The stimuli subtended 8.5° from the center of the screen, with the outer fixation points being 11° from the center. (Reproduced from Tovée MJ, Rolls ET, Azzopardi A. J Neurophysiol 1994; 72:1049-1060. © (1994) The American Physiological Society).

fixation positions was 2.3 bits. Thus, the cells could, if they had been responsive to position, have provided more information about fixation position than about identity.

For identity the bulk of the information was coded by the first principal component (which is highly correlated with the firing rate). Little information was added by the addition of higher principal components. The situation is complementary for fixation position. The first principal component carried comparatively little information about fixation position, with most information being carried by the second principal component. The latency for the neuronal responses when the fixation position is away from the center of the face was 20-30 msec longer than for fixation on the center of the face, and this latency difference was reflected in the second principal component.[7] This finding will be further discussed in chapter 3.

To determine whether the centers of the cells' receptive fields were systematically offset with respect to each other (implying that there might be a distributed representation of information about where a stimulus was with respect to the fovea), the center of gravity of the responses of each neuron were calculated as:

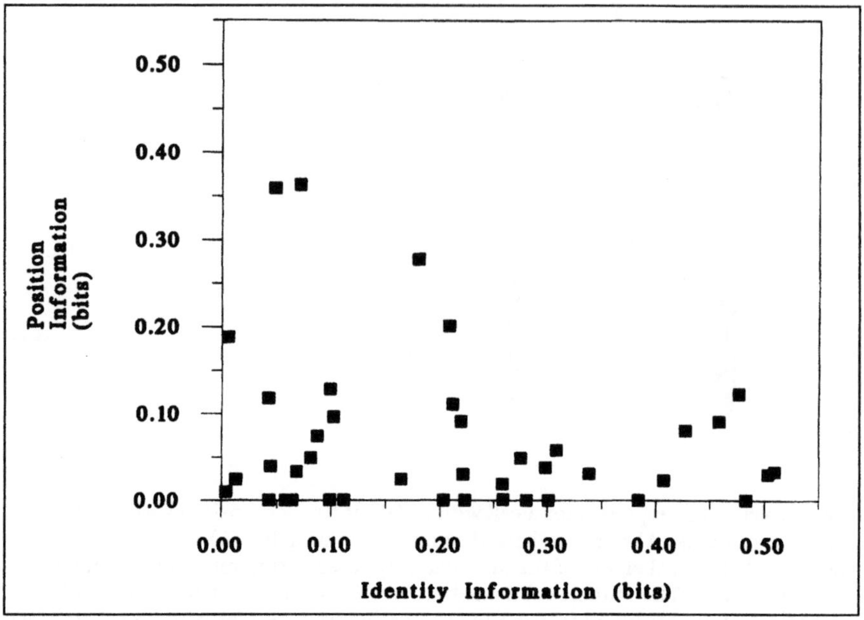

Fig. 2.2. Information contained in the neuronal response of each of 44 cells about which stimulus had been shown (abscissa), and about the fixation position on the stimulus (ordinate) for the different cells. Each point shows the data from a single cell. The information shown was based on principal component analysis of a 400 msec period of the neuronal spike train starting 0 msec after the stimulus appeared. The information was calculated from the first three principal components. (Reproduced from Tovée MJ, Rolls ET, Azzopardi A. J Neurophysiol 1994; 72:1049-1060. © (1994) The American Physiological Society).

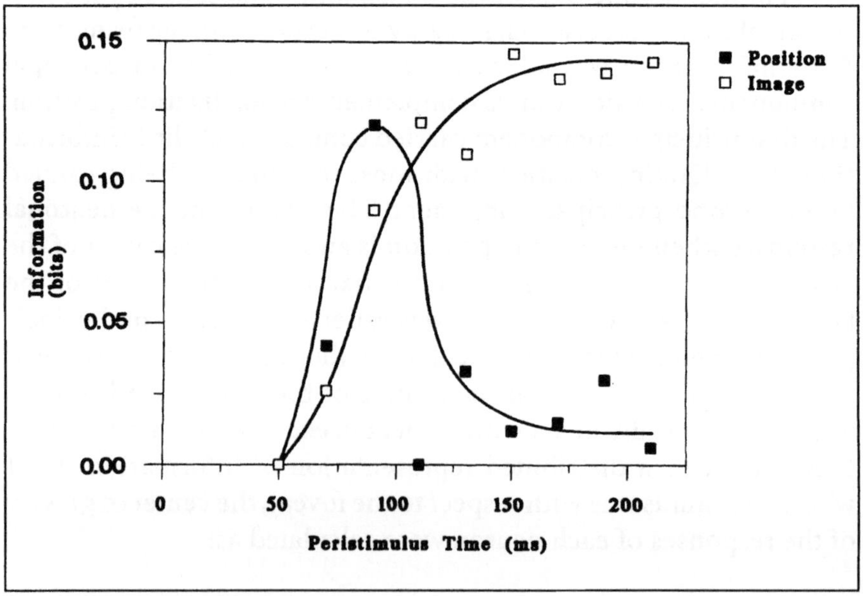

$$x = \langle f_i \bullet x_i \rangle / \langle f_i \rangle$$

$$y = \langle f_i \bullet y_i \rangle / \langle f_i \rangle$$

where *x* and *y* are the x and y coordinates of the center of gravity, f_i is the firing rate at position x_i, y_i, and $\langle \cdot \rangle$ indicates the ensemble average.[5]

The distances of the centers of gravity from the fovea of the different cells are shown in Figure 2.4, scaled so that a distance of 1 unit corresponds to 11° The centers of gravity were grouped relatively close to the fovea, with the great majority having centers of gravity within 3° of the fovea.

If the cells recorded from in this study have bilateral responsive fields, receiving inputs from both the contralateral and ipsilateral visual fields, then the ratio of the responses to stimuli on the left side to stimuli on the right side of the fovea should be one. This comparison is shown in Figure 2.5. Each square represents the response of a single cell. Cells with equal responses in the two visual fields would fall along the dashed line. The majority of the cells did not significantly differ in position from this line.

This study shows that there is a population of neurons in anterior IT, with large, bilateral receptive fields which show translation invariance. This finding is consistent with several other studies which have suggested that stimulus selectivity is largely invariant for changes in stimulus position within the receptive field.[8-14] Moreover, it seems that in general the shape and color selectivity of IT cells is invariant for changes in the size of the stimulus.[8-14]

Objects need to segregated from their background to be recognized. Many perceptual studies have shown that the discontinuities in the distribution of several parameters lead to a salient segregation of figure and background (e. g., discontinuities in luminance, depth, color, texture and relative motion). What this means is that the same object can be segregated from its background using several visual cues and can be identified as that object irrespective of its defining cue. This cue invariance feature of object recognition is obviously important, as the same object may be recognized under a variety of different conditions, in some of which it will stand out

Fig. 2.3 (opposite). Information that is available in 20-msec time bins of the firing rate about fixation position (filled squares) and image identity (open squares) for the population of 44 squares analyzed. (Reproduced from Tovée MJ, Rolls ET, Azzopardi A. J Neurophysiol 1994; 72:1049-1060. © (1994) The American Physiological Society).

Fig. 2.4. Center of gravity of the receptive field of each of the cells analyzed, together with its standard deviation estimated from the standard deviation of the firing rate. (Reproduced from Tovée MJ, Rolls ET, Azzopardi A. J Neurophysiol 1994;72:1049-1060. © (1994) The American Physiological Society).

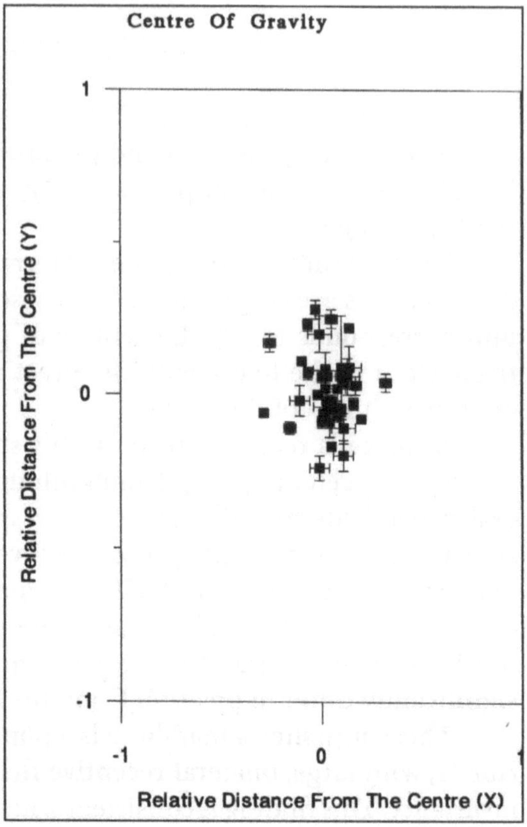

Fig. 2.5 (below). Magnitude of the responses of 44 cells to full-sized stimuli shown contralateral (on the left side of the visual field) or ipsilateral (on the right side of the visual field) to the recording site in the right hemisphere. Each square represents the response of a single cell. Cells with equal responses would appear along the dashed line. The monkey fixated 11° from the center of the stimuli and the stimuli extended 8.5° from the center. (Reproduced from Tovée MJ, Rolls ET, Azzopardi A. J Neurophysiol 1994;72:1049-1060. © (1994) The American Physiological Society).

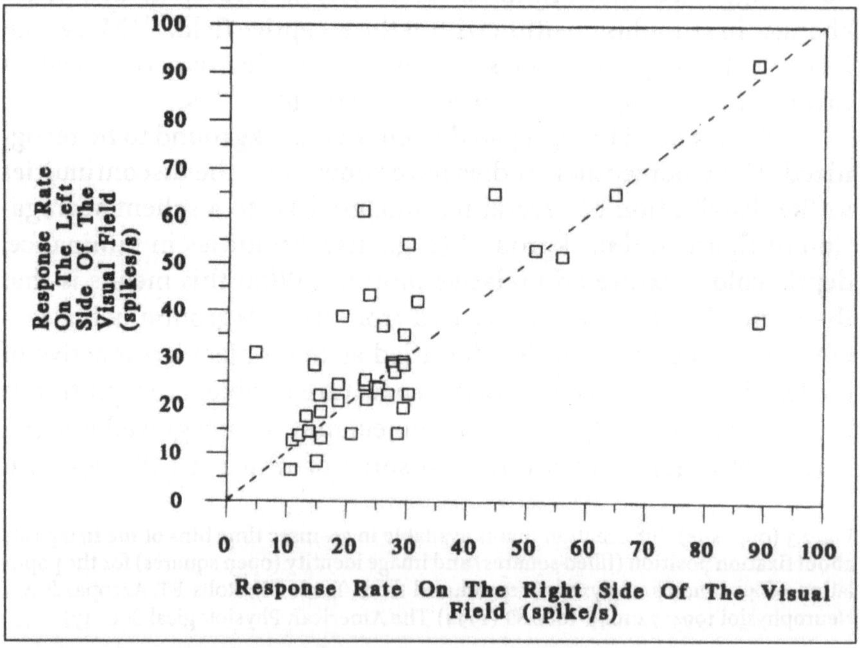

from its background by virtue of only a single visual cue. To explore this theme, Guy Orban's group recorded from visual neurons in the lower bank of the superior temporal sulcus (STS), close to the lateral convexity, or in the lateral part of the IT itself. They found neurons which responded to shapes defined by one of several possible cues, such as luminance, texture or motion.[10,15]

However, it is not only invariance to the cues that define an object that must be developed by the visual system. In our normal environment, many objects are often partly occluded by other objects. Yet despite their partial occlusion, we are able to recognize these objects. Object recognition seems to be invariant with respect to change in the retinal image due to partial occlusion of the object.[16-17] Further recording in IT cortex has shown a population of IT neurons that are responsive to partially occluded shapes, and that the stimulus selectivity of these neurons is similar to that for unoccluded shapes.[16] However, the protocol used in this experiment casts some doubt on the validity of this result. There was a time lag between the presentation of the occluding pattern and the occluded shape. This time lag could act to enhance the perceptual segregation of the shape and the occluder. The result might have been different for simultaneous presentation of occluder and shape. However, these results do show that under some conditions, under which monkeys were able to discriminate partially occluded shapes, IT cells seemed to remain shape selective. One should, however, also point out that this is not the same as amodal completion. Under conditions of partial occlusion, humans (and probably monkeys) perceive the partially occluded shapes as single complete forms (amodal completion). This has not been demonstrated at the single IT cell level.[16-17] Cells responded on average as well to shape parts that were perceived as isolated, disconnected units as they did to shape parts that, to humans, were seen as part of a gestalt whole. It is possible that neurons higher up the processing stream (more anterior in IT for example) might have displayed amodal completion, but at this time there is no known unequivocal basis for this perceptual phenomenon.

The different forms of response invariances observed in IT emerge gradually within the visual system. As we move up the object processing pathway in monkeys (V1-V2-V4-posterior IT-anterior IT), the response properties of the neurons change. The receptive field of a cell gets significantly larger. For example, the average receptive field size in V4 is 4 degrees^2 which increases to 16 degrees^2 in posterior IT, and to 150 degrees^2 in anterior IT. Most cells along the V4, posterior IT and anterior IT pathway also have receptive fields close to, or including, the fovea (75% of anterior IT cells included the fovea). The increase in receptive field allows the development of a visual response that is unaffected by the size and position of a stimulus within the visual field. Some features of visual cue invariance similar to those observed in it have also been reported in earlier visual areas, for example, illusory contours in V1,[18] illusory contours and stereo and kinetic boundaries in V2[19-21] and kinetic boundaries in V4.[22] This neurophysiological evidence is supported by a psychophysical study by Davis and Driver,[23] who used a visual search task to distinguish between early and late stages in the processing of visual information. For example, among many jumbled white letters a single red one is instantly discerned (a phenomenon called "pop out"), but a single L among many Ts needs more time to be detected. This result is taken to suggest that color differences are extracted early in the visual system, but differentiation of similar letters is the result of more complex processing at a higher level. This procedure can be quantified by measuring the time it takes for a single odd feature to be detected among a number of background features. A rapid reaction time which is largely independent of the number of background features is taken to be indicative of processing at an early stage in the visual system. Davis and Driver used figures outlined by illusory contours based on the Kanizsa triangles, and their results were consistent with the processing of these features occurring early in the visual system.

Taken together, these results suggest that relatively early in the visual system, several visual cues already converge on single neurons. These neurons can signal the presence of a boundary irrespective of the defining cue. Convergence of information from multiple cues helps distinguish real boundaries from accidental, noise induced signals, since discontinuities in many cues (luminance, depth, color etc.) usually coincide spatially for real, but not noise, boundaries. This could explain why cue convergence is found at early stages of the visual system. The IT neurons showing cue invariance are likely to receive input from the invariant "boundary" neurons of lower visual areas.[17]

The gradual concentration of information in the higher visual areas of the ventral system has been elegantly demonstrated using the technique of binocular rivalry. If different images are presented to the left and the right eye, then the subject does not perceive a combination of the two images, but each image alternately. Using a monkey who has been trained to make differential reports according to two predetermined stimulus categories, it was possible to record in ascending visual areas to see how well correlated neuronal responses are with perception. Only 18% of the cells in V1 responded in the same way as the whole animal did, but this increased to 25% in V4 and almost 100% for IT neurons.[24-25] This suggests that conscious perception can reliably be correlated with neuronal activity only in IT and supports the proposition that visual awareness is mediated specifically by the higher visual areas.[26-27]

How Are Stimuli Represented in the Temporal Visual Cortex?

Tanaka has tried to determine the minimum features necessary to excite a cell in anterior IT using a procedure which simplifies the stimuli used to excite a given neuron.[28-29] This procedure begins by presenting a large number of patterns or objects while recording from the neuron, to find which objects excite the cell. Then the component features of the effective stimulus are segregated and presented singly or in combination, while assessing the strength of the cell's response for each of the simplified stimuli. The aim is to find the simplest combination of stimulus features to which the cell maximally responds. However, feature space is a vast multi-dimensional area in which even the simplest "real world" stimulus will possess a wide variety of elementary features, such as depth, color, shape, orientation, curvature and texture and may show spectacular reflections and shading.[30-31] It is therefore not possible to present all the possible feature combinations systematically, and the simplified stimuli that are actually presented in the cell's receptive field are typically a subset of the possible combinations. Hence, it is not possible to conclude that the best simplified stimulus is optimal for the cell, only that it was the best of those presented.

Tanaka found a population of neurons in IT, which he called elaborate cells, that seemed to be responsive to simple shapes.[32-33] Moreover, Tanaka found that closely adjacent cells usually responded to very similar feature configurations. In vertical penetrations through the cortex, he consistently recorded cells that responded to the same "optimal" stimulus as for the first test cell tested, indicating that cells with similar preferences extend through most cortical layers. In tangential penetrations, cells with similar preferences were found in patches of approximately 0.5 mm^2. These results suggest that the cells in IT are organized into functional columns or modules, each module specifying a different type of shape. If these modules are 0.5 mm^2 in width, then there could be up to up to 2,000 within IT. However, allowing for the fact that many may analyze the same type of shapes, and many may analyze more complex patterns such as faces, the number of different simple shapes is probably only around 600.[34]

In a recent study combining optical imaging with electrophysiology, a patchy distribution of staining was shown, which might be consistent with a columnar organization.[35] However, the staining also showed that instead of discreet columns with little overlap, there was actually significant overlap. This has led to the suggestion that stimuli could be mapped as a continuum of changing features.[29] Such a continuous map could produce a broad tuning of cortical cells for certain directions of feature space, which would allow the association of different, but related, images, such as the same object from different viewpoints or under different illumination.[29] This would obviously be an important mechanism in the development of a stimulus-invariant response. However, as mentioned above, feature space is multi-dimensional and so a continuous representation would have to be reduced in some way to fit the limited dimensions possible in the cortex. Ultimately, a columnar organization is more likely, with cells in several columns responsive to stimuli which have features in common, becoming jointly active—a scheme which can also give rise to stimulus invariance (Fig. 2.6).

It has been suggested that the simple-shape preferences of these cells may form a "visual alphabet" from which a representation of an object can be constructed.[36-37] The number of these simple shapes is very small by comparison to the number of possible visual patterns, in the same way that the number of words that can be constructed from an alphabet is very large. Each cell would signal the presence of a particular simple shape if it was present in a complex pattern or object. Based on the responses of these elaborate cells, a representation could be derived in at least two ways. Firstly, there could be a traditional hierarchy in which these elaborate cells would feed into a higher cell layer, whose cells respond preferentially to complex stimuli. The output of these cells would then signal the presence of a complex object to higher areas, such as the prefrontal cortex. Alternatively, there may be no upper layer. The pattern of responses across

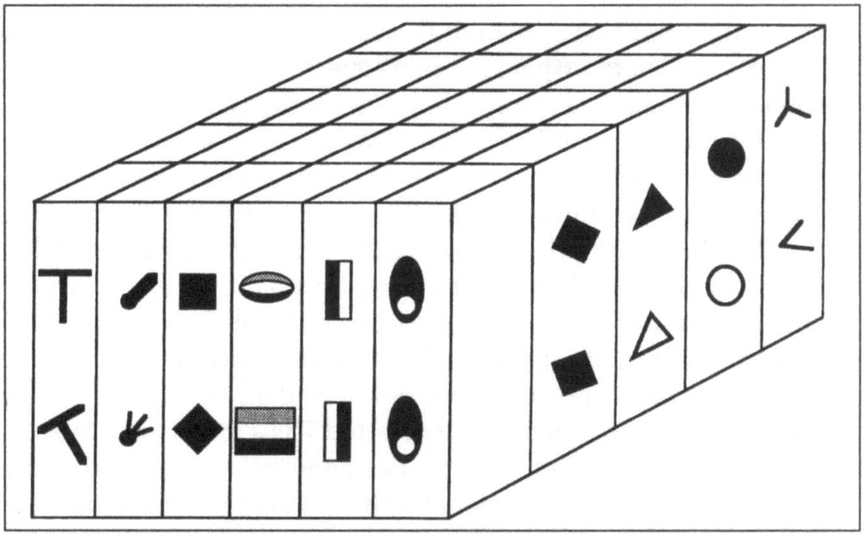

Fig. 2.6. Schematic diagram of the columnar organization of the inferior temporal cortex. The average size of columns across the cortical surface is 0.5 mm. Cells in a particular column have similar but slightly different selectivities. (Redrawn from Tanaka K. Curr Op Neurobiol 1992: 2: 502-505).

the various columns of elaborate cells may directly signal the presence of a complex object to a higher area, without having to converge on a cell in IT sensitive to complex stimuli.

The concept of the visual alphabet assumes that an IT cell will reliably signal the presence of the particular simple shape that excites it, regardless of whatever else is present in the visual field. However, this may not always be the case.[38] In a recent study, the responses of single neurons in the anterior IT of awake behaving macaques were recorded when the monkey performed a visual fixation task.[39] The responses of individual neurons to visual stimuli presented individually were then determined, at one of two positions in the cell's receptive field. The images were then shown in pairs, and the cell's response compared to its response to the images when they were presented separately in the corresponding area of the receptive field. The results showed a significant interaction effect between the stimuli when shown as pairs (Fig. 2.7). A similar effect has been reported for neurons in more posterior IT.[40-41] Additionally, some of Tanaka's own data may show this effect. In the example shown in Figure 2.8, Tanaka's simplification procedure converged on an inverted T-shape as the preferred simple shape for this cell.[32] A more complex object that contains this simple shape should evoke a strong response from the cell, as the cell should signal the presence of the shape. Just such an example is present in figure, together with the cell's response to it. The cell did not respond well to a + shape, in which the preferred simple shape is still present, in concert with a bar below its center. Thus, the presence of other visual features can disrupt the response of a cell to its preferred shape, a result that is the opposite of that assumed in the visual alphabet conception of IT. The cell may be signaling a combination of the presence of its preferred shape and the absence of something else.[30] The characteristics of what must be absent for the cell to respond have not been defined, except to show that a bar below the center of the inverted T is a member of this set, and so the simplification protocol has gone only half-way towards defining the conditions sufficient for the cell's response. If other cells behave in a similar way, the characterization by this method of the simple shapes preferred by IT cells cannot be sufficient to account for the performance of the cells in the recognition of even slightly more complex objects.

Face-Selective Neurons

There is evidence that the cellular coding of at least some complex patterns and objects does not remain as a collection of separate codes for its component shapes. The most studied example is the face-selective neuron. For more than 25 years it has been known that there are neurons in the monkey temporal visual cortex which are sensitive to faces, hands and other complex biological stimuli.[42-43] Those responsive to faces are localized in the inferior temporal cortex, and the banks and walls of the superior temporal sulcus (STS). The optimal stimuli of a large proportion of these cells cannot be deconstructed into simpler component shapes.

In general, cells that are responsive to faces show virtually no response to any other stimulus tested (such as textures, gratings, bars, and the edges of various colors) but respond strongly to a variety of faces, including real ones, plastic models, and video display unit (VDU) images of human and monkey faces.[44-46] The responses of many face-selective neurons are size and position invariant; the cell's response is maintained when there is a 12-fold change in the size of the face,[47] or if the position of the face within the cell's receptive field is altered.[5,45] Some face cells do not respond well to VDU images of faces that have had the components rearranged, even though

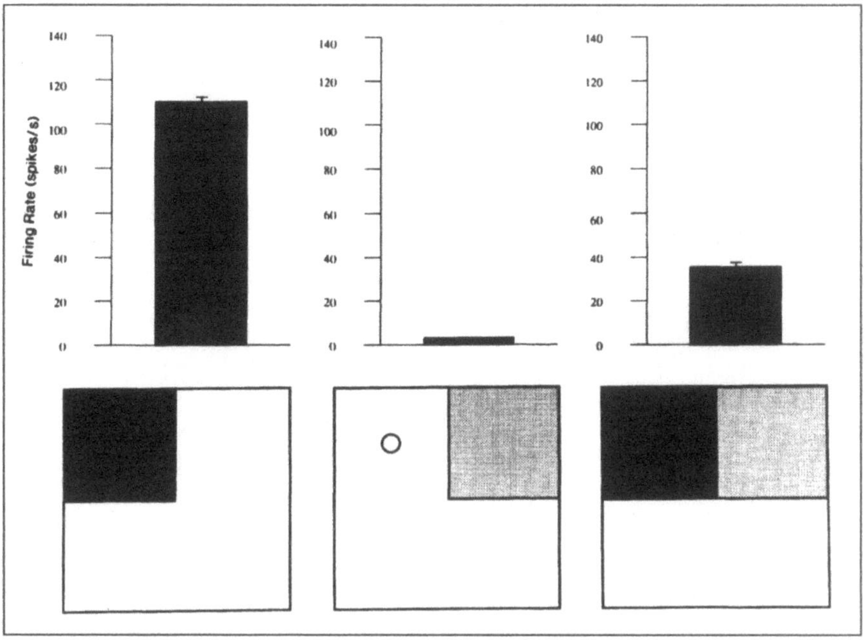

Fig. 2.7. The neuronal firing rate of one cell when an effective stimulus was fixated and a noneffective stimulus was presented para-foveally. Fixation was always at the point shown by the small ring in the upper left-hand quadrant. *Left*, the neuron had a large firing rate when the effective stimulus was at the fixation point. *Middle*, the neuron did not respond when the noneffective stimulus was presented para-foveally. *Right*, the firing rate of the neuron to the effective stimulus at the fixation point was significantly reduced by the noneffective stimulus presented para-foveally. The means and the standard errors are shown. The cell's spontaneous firing rate was 10 spikes/s. (Reproduced with permission from Rolls ET, Tovée MJ, Exp Brain Res 1995; 103:409-420. © Springer-Verlag).

all the components are still present and the outline is unchanged.[45-46,48] Face-selective neurons are sensitive to the relative position of features within the face; particularly important is intereye distance, distance from eyes to mouth, and the amount and style of hair on the forehead.[49-50] Moreover, presentation of a single facial component elicits only a fraction of the response generated by the whole face, and removal of a single component of a face reduces, but does not eliminate, the response of a cell to a face.[45-46] Face-selective neurons also continue to respond to VDU images of faces that have been low or high-pass band filtered so that they have no spatial frequencies in common,[51] that have had the color removed or altered,[46-47] or have had the contrast reduced to a very low level.[47] Line drawings of faces produce a weak response.[44,46] These complex

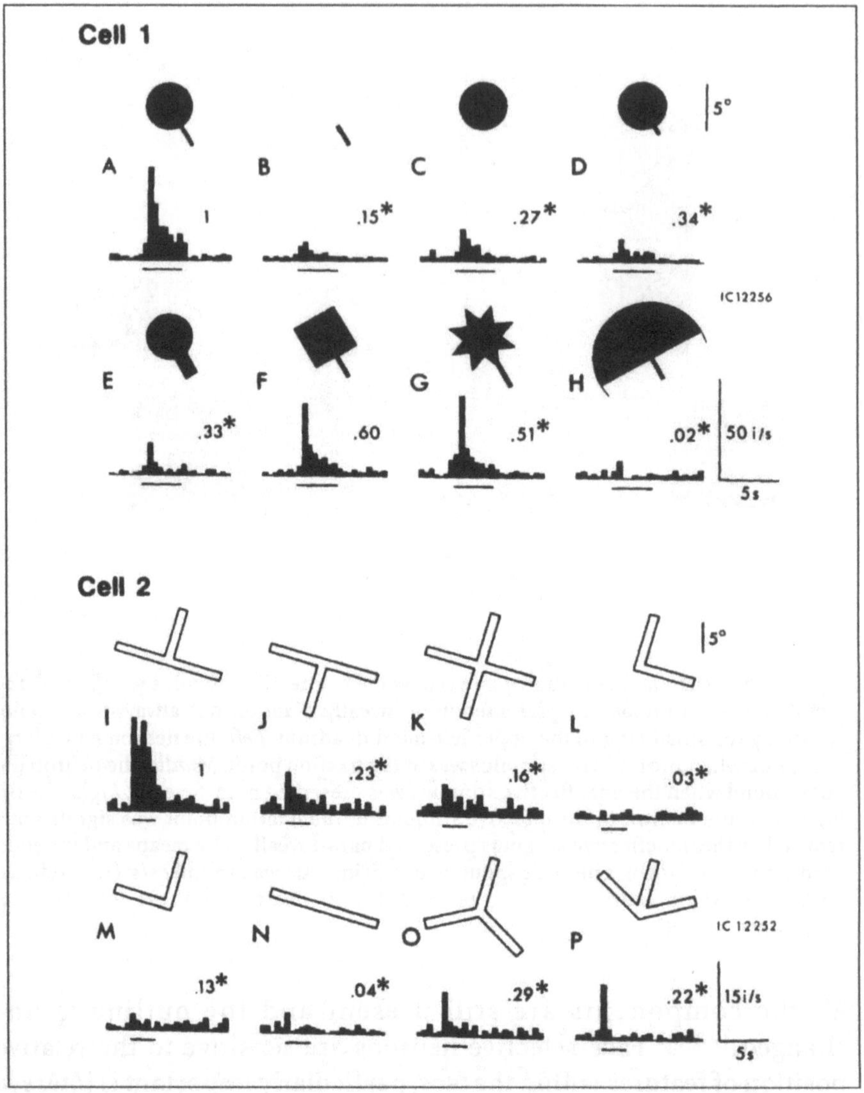

Fig. 2.8. An example of the responses of a two separate cells to a number of stimuli used in Tanaka's simplification procedure. For Cell 1 the simplification procedure converged on a black circle and attached bar (Stimulus A). For Cell 2 the simplification procedure converged on an inverted T shape (Stimulus A). If the visual alphabet theory is correct, then a cell should respond strongly to any more complex object that contains the optimal stimulus shape as determined by the simplification procedure. For Cell 2, a more complex object containing this shape is stimulus C. The response of the cell to this stimulus is very weak, which contradicts the visual alphabet theory. (Reproduced with permission from Tanaka K, Saito H, Fukado Y, Moriya M, J Neurophysiology 1991; 66:170-189. © (1991) The American Physiological Society).

neuronal properties suggest that this class of cell is truly face selective, and not responding to some other characteristic of the visual image.

The response of the majority of cells in the STS and IT are view-selective, but their outputs could be combined in a hierarchical manner to produce view-independent cells in anterior IT.[52] If this were true, then the response latency of view-independent cells should be longer than that for view-dependent cells, which seems to be the case. The mean latency of view-independent cells (130 msec) was significantly longer than that for view-dependent cells (119 msec).[43]

Combined electrophysiological and tracing studies suggest that face-selective neurons responsive to a particular aspect of a face, such as the angle at which the head is observed, seem to be found grouped together in chords or mini-columns that run perpendicular to the cortical surface.[53-54] This finding was confirmed by a combined electrophysiological and optical imaging study using face stimuli.[29,35] The preferred stimulus for individual cells were first determined in single unit recordings with microelectrodes. In the subsequent optical imaging study, presentation of the preferred face stimulus activated patchy regions around 0.5 mm in diameter, covering the site of the electrode penetration at which the face-selective neuron had been located. Changing features of the face, such as rotating the face from profile to full-face produced a systematic movement in the activated patch (Fig. 2.9). These results are consistent with a columnar organization of face-selective neurons, with neurons with similar stimulus selectivity's being grouped together. However, as seen from Figure 2.9, there is some overlap in activity between different cell populations. This suggests joint activation of cells in different columns that are responding to features which different views of the rotated face have in common. This allows the development of stimulus-invariance, while the changing pattern of activation will also signal changes in stimulus orientation.

What Do Face-Selective Neurons Do?

Face cells within the macaque temporal lobes have been reported within both the inferior temporal gyrus and the banks and floor of the superior temporal sulcus. The inferior temporal gyrus and the lower banks of the sulcus are part of IT cortex,[42,55] whereas the upper bank of the sulcus contains a high proportion of polysensory neurons and is called the superior temporal polysensory

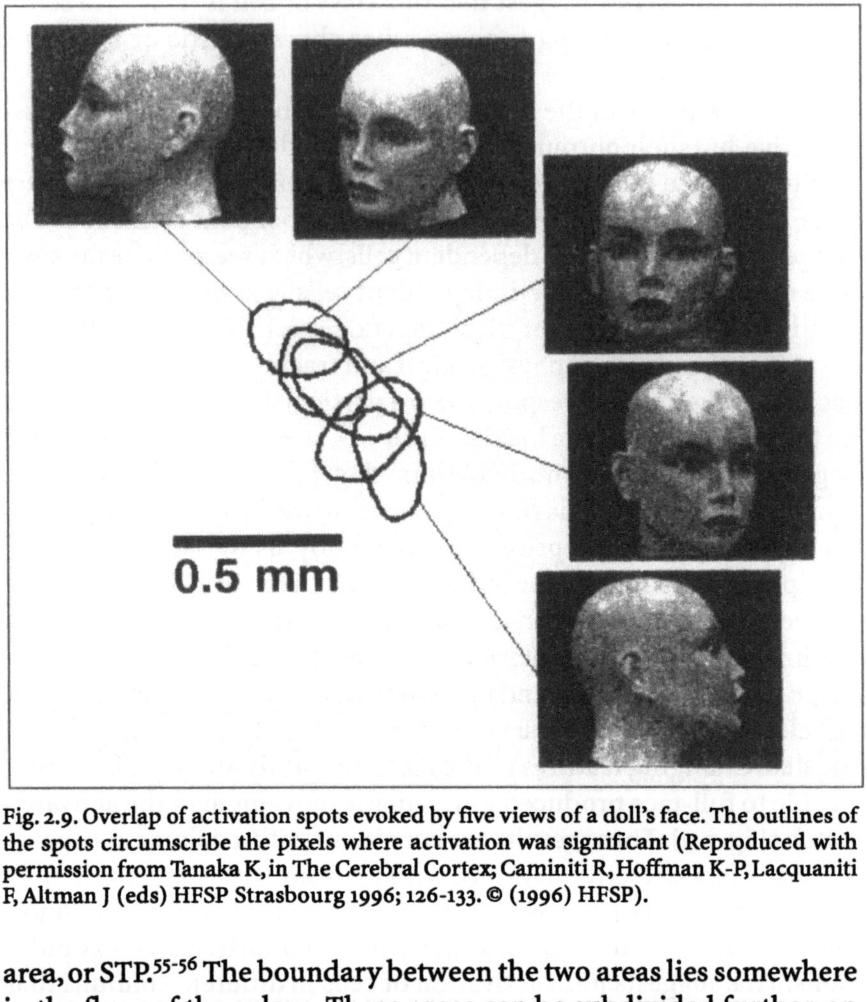

Fig. 2.9. Overlap of activation spots evoked by five views of a doll's face. The outlines of the spots circumscribe the pixels where activation was significant (Reproduced with permission from Tanaka K, in The Cerebral Cortex; Caminiti R, Hoffman K-P, Lacquaniti F, Altman J (eds) HFSP Strasbourg 1996; 126-133. © (1996) HFSP).

area, or STP.[55-56] The boundary between the two areas lies somewhere in the floor of the sulcus. These areas can be subdivided further, on the basis of cyto-architecture and anatomical connections, into six subdivisions; areas TAa, TPO and PGa are located within the STP, areas TEm and TEa are located within the IT cortex, and area IPa is located on the interface between the STP and IT cortex[57] (Fig. 2.10). Baylis et al[56] reported that face cells were found throughout all divisions of the sulcus and the inferior temporal gyrus, but were especially heavily concentrated in the upper bank of the superior temporal sulcus (within area TPO of the STP) and in the lower bank and lip of the sulcus (within area TEm of the IT cortex).

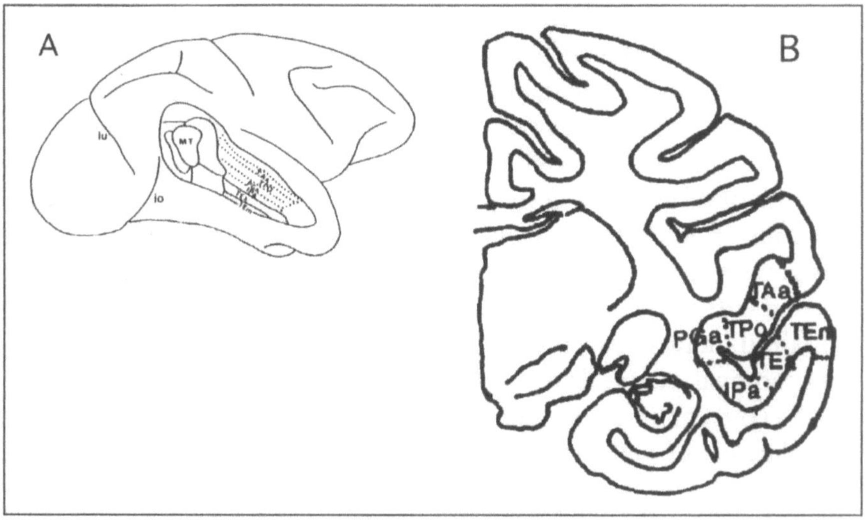

Fig. 2.10. Lateral and coronal views of the macaque brain showing the superior temporal sulcus (STS), which has been opened out to display the subdivisions proposed by Seltzer and Pandya.[57] Areas TAa, TPO, and PGa comprise the superior temporal polysensory area (STP) and areas TEm and TEa comprise the inferior temporal cortex (IT). IPa is located at a position intermediate between the STP and IT. Sulcal abbreviations: lu = lunate sulcus, io = inferior occipital sulcus (Reproduced with permission from Tovée MJ, Cohen-Tovée EM, Cog Neuropsychol 1993; 10: 505-528).

The two different regions have very different physiological properties and anatomical connections. IT neurons are usually selective for object features such as shape and color, and have large receptive fields that are most sensitive at the fovea. However, the STP cells, except for the face cells, are rarely sensitive to shape or color, and their receptive fields are at least as sensitive to the far periphery of the visual field as to the fovea.[44] Areas in the STP corresponding to the regions where face cells are found receive strong projections from the inferior parietal lobe, which is part of the dorsal processing stream involved in the processing of spatial information.[57-58] The STP also receives projections from the medial superior temporal and fundus of the superior temporal visual areas,[59] which are involved in the processing of visual motion information. This is consistent with the responsiveness of many STP cells to movement,[44] including many which are preferentially responsive to faces or to body posture.[52,60]

Hasselmo et al[61] measured the responses of face cells to 9 pictures of 3 different monkey faces, each with 3 different facial expressions (calm, slight threat, or full threat). Thirty-three per cent of the cells showed significant selectivity for identity independent of expression, and 20% showed significant selectivity for expression independent of identity. Only 7% of the cells showed a significant interaction effect. Cells sensitive to identity were located on the inferior temporal gyrus (IT cortex) and those sensitive to expression were located within the superior temporal sulcus (STP). Monkeys are not impaired on face discrimination and identification tasks after bilateral removal of the floor and banks of the superior temporal sulcus.[62] Thus, expression and identity seem to be coded by separate populations of cells, located in adjacent but anatomically discrete regions.

The face-selective neuron population in STP can be subdivided further into two or more functional groups. In addition to cells responsive to the emotion a face is displaying, there is evidence to suggest that there are also cells responsive to where in the environment other individuals are directing their attention. Identifying the direction of attention of other individuals within a primate social group is extremely important. It is not enough to know that an individual has a threatening expression; it is also necessary to know at whom the expression is directed. The majority of face cells in the STP are responsive to the direction in which the head points,[53] which is consistent with the strong input from the inferior parietal lobe. Cells selectively responsive to a particular view have been identified across the entire 360° range of views,[4,54,60] although Perrett et al[60] suggest that these cells show a statistical preference for face, profile, and rear view compared to intermediate views. The direction in which the head points may provide an index of where the attention of other individuals is directed.[60] It has also been shown that the response of a cell to a particular view of a head can be affected by the head's direction of gaze.[52,60] Moreover, bilateral removal of the banks and floor of STS severely impairs the ability to perceive the gaze direction of a face.[61-62] Harries and Perrett[60] have suggested that analysis of information about the direction of attention may use a system of defaults such that if the eyes are visible, then processing is guided by gaze direction. For those views of the head where gaze is not visible the direction in which the head is pointing is used as an index of the direction of attention.

This evidence is complemented by the results of anatomical studies which show that the STP (the upper bank of STS, TPO, and PGa) receive strong projections from the caudal third of the inferior parietal lobe.[57-58,63-64] The pattern of the parietal projections to the STP corresponds to the modular distribution of the direction of attention-selective cells;[53] there are also strong projections from these temporal areas to the parietal lobe.[53] The tempero-parietal projections could provide a means by which information could be exchanged between the ventral and dorsal processing streams, so that information from the temporal lobe analysis of facial signals about the direction of others' attention can be passed to the parietal systems concerned with spatial processing, and vice versa.

Cells in anterior STP also seem to be responsive to "biological motion".[65-66] These neurons are responsive to whole body movements, such as walking, and are also responsive to the direction of motion and the angle at which the body is observed.[66-67] This is again consistent with the integration of motion and 3 dimensional information with object information in STP.

Are Faces Special?

Tanaka[29,68] has argued that faces are "special". He postulates that all other complex stimuli are represented by a distributed code across all the "elaborate cells' of the visual alphabet. Only faces have a specific class of neurons tuned to their representation (see Fig. 2.11). However, this may not be a tenable position. It seems unlikely that only faces, out of all the complex stimuli that the brain needs to represent, should have a specific class of cells to represent them. Faces are important, especially for social animals such as primates, but so are other visual stimuli, such as food sources and predators. It seems more likely that face-selective neurons are an example of a class of neurons responsive to complex stimuli, and that other classes of neurons are responsive to other classes of complex stimuli. This is consistent with the findings of single-cell recording studies which have consistently reported IT cells responsive to other complex biological stimuli, such as hands,[42,69] and to complex 3 dimensional shapes.[13] This is also supported by clinical data from brain damaged patients, who have reported selective impairments not only of face recognition, but of the ability to recognize and classify complex and patterns, such as coins, cars or domestic animals.[70]

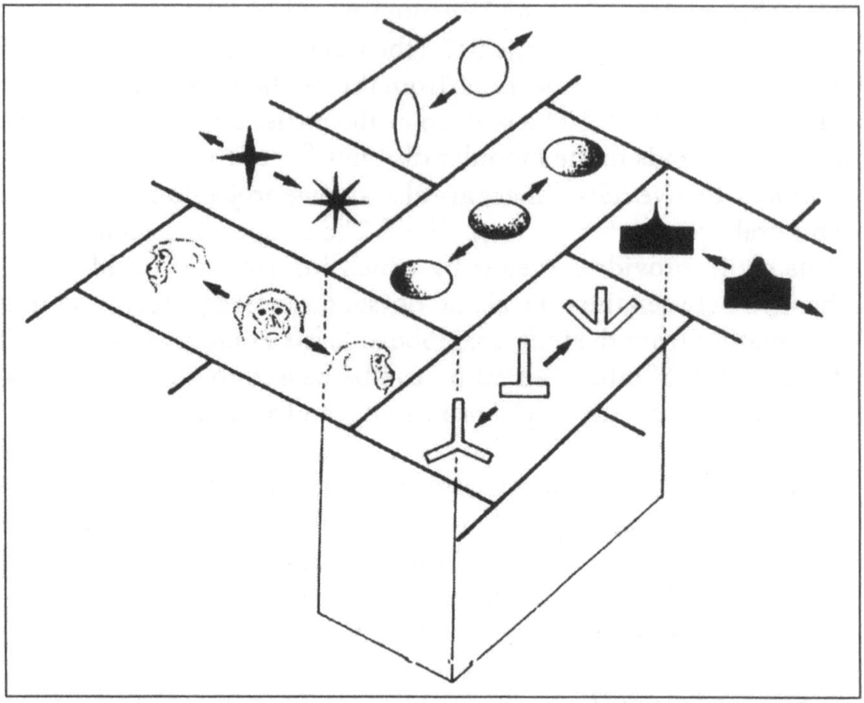

Fig. 2.11. Modification of the schematic diagram of the columnar organization of the inferior temporal cortex to incorporate face-selective neurons. (Redrawn from Tanaka K, in The Cerebral Cortex; Caminiti R, Hoffman K-P, Lacquaniti F, Altman J (eds) HFSP Strasbourg 1996; 126-133.).

Face processing is probably no different from the visual processing of any other complex object within the ventral processing stream.[71-72] However, the processing of face information can be said to be special in terms of the amount of cortex devoted to its analysis and the complexity of connections in this region of cortex. This reflects the importance of facial information as a medium of communication to highly sociable primates such as man.

Primates exist in complex, hierarchical, social groups and face recognition is therefore important for identifying other members of the group (in order to adopt the right stance towards them on the basis of their social position within the group) and for communication via facial expression. It has been suggested that the importance of face recognition among social animals has exerted a selective pressure that has led to the evolution of neural mechanisms for the analysis of facial information.[69,72] In primates this development has evolved in tandem with a greater differentiation of facial musculature, which is capable of more complex facial expressions. Neural mechanisms selective for faces are at their most complex in man, where social organization is at its most advanced and facial musculature is at its most complex.

References

1. Marr D. Vision: A computational investigation into the human representation and processing of visual information. San Francisco: W H Freeman and Co 1982.
2. Weiskrantz L. Visual prototypes memory and the inferotemporal cortex. In: Iwai E, Mishkin M eds. Vision memory and the temporal lobe. New York: Elsevier 1990.
3. Boussaoud DB, Desimone R, Ungerleider LG. Visual topography of area TEO in the macaque. J Comp Neurol 1991; 306:554 -575.
4. Hasselmo ME, Rolls ET, Baylis OC. The role of expression and identity in the face-selective responses of neurons in the temporal visual cortex of monkey. Behav Brain Res 1989; 32:203-218.
5. Tovée MJ, Rolls ET, Azzopardi A. Translation invariance in the responses to faces of single neurons in the temporal visual cortical areas of the alert macaque. J Neurophysiol 1994; 72:1049-1060.
6. Optican LM, Gawne TJ, Richmond BJ, Joseph PJ. Unbiased measures of transmitted information and channel capacity from multivariate neuronal data. Biol Cybern 1991; 65:305-310.
7. Tovée MJ, Rolls ET, Treves A, Bellis RP. Information encoding and the responses of single neurons in the primate temporal visual cortex. J Neurophysiol 1993; 70:640-654.
8. Schwartz EL, Desimone R, Albright TD, Gross CG. Shape recognition and inferior temporal neurons. Proc Natl Acad Sci USA1983; 80:5776-5778.
9. Komatsu H, Ideura Y, Kaji S, Yamane S. Color selectivity of neurons in the inferior temporal cortex of the awake macaque monkey. J Neurosci 1992; 12:408-424.
10. Sary G, Vogels R, Orban GA. Cue-invariant shape selectivity of macaque inferior temporal neurons. Science 1993; 260:995-997.
11. Lueshow A, Miller EK, Desimone R. Inferior temporal mechanisms for invariant object recognition. Cereb Cortex 1994; 5:523-531.
12. Ito M, Tamura H, Fujita I, Tanaka K. Size and position invariance of neuronal responses in monkey inferior temporal cortex. J Neurophysiol 1995; 73:218-226.
13. Logothetis NK, Pauls J, Poggio T. Shape representation in the inferior temporal cortex of monkeys. Current Biology 1995; 5:552-563.
14. Sato T, Kawamura T, Iwai E. Responsiveness of inferotemporal single units to visual pattern stimuli in monkeys performing discrimination. Exp Brain Res 1980; 38:313-319.

15. Sary G, Vogels R, Kovacs G, Orban GA. Responses of monkey inferior temporal neurons to luminance- motion- and texture-defined gratings. J Neurophysiol 1995; 73:1341-1354
16. Kovacs G, Vogels R, Orban GA. Selectivity of the macaque inferior temporal neurons for partially occluded shapes. J Neurosci 1995; 15:1984-1997.
17. Vogel R, Orban GA. Activity of inferior temporal neurons during orientation discrimination with successively presented gratings. J Neurophysiol 1994; 71:1428-1451.
18. Grosof DH, Shapley RM, Hawken MJ. Macaque V1 neurons can signal 'illusory' contours. Nature 1993; 365:550-552.
19. Von der Heydt R, Peterhans E, Baumgartner G. Illusory contours and cortical neuron responses. Science 1984; 224:1260-1262.
20. Von der Heydt R, Zhou H, Friedman H, Poggio GF. Neurons of area V2 of visual cortex detect edges in random dot stereograms. Soc Neurosci Abstr 1995; 21:18.
21. Marcar VL, Xiao DK, Raiguel SE, Orban GE. Selectivity of area V2 of the macaque to kinetic and other types of boundaries. Soc Neurosci Abstr 1994; 20:1740.
22. Logothetis NK, Charles ER. V4 responses to gratings defined by random textured motion. Invest Ophthalmol Visual Sci 1990; 31:44.
23. Davis G, Driver J. Parallel detection of Kanizsa subjective figures in the human visual system. Nature 1994; 371:791-793.
24. Leopold DA, Logothetis NK. Activity changes in early visual cortex reflect monkey's percepts during binocular rivalry. Nature 1996; 379:549-553.
25. Sheinberg DL, Logothetis NK. The role of temporal cortical areas in perceptual organization. Proc Natl Acad Sci USA 1997; 94:3408-3413.
26. Crick F, Koch C. Are we aware of neural activity in primary visual cortex? Nature 1995; 375:121-123.
27. Koch C, Braun J. Towards the neural correlates of visual awareness. Curr Op Neurobiol 1996; 6:158-164.
28. Kobatke E, Tanaka K. Neuronal selectivities to complex object features in the ventral visual pathway of the macaque cerebral cortex. J Neurophysiol 1994:71:856-867.
29. Tanaka K. Inferotemporal cortex and object recognition. In: Caminiti R, Hoffman K-P, Lacquaniti F, Altman J eds. Vision and Movement Mechanisms in the Cerebral Cortex. Strasbourg: HFSP 1996; 126-133.
30. Young MP. Open questions about the neural mechanisms of visual pattern recognition. In: Gazzanida MS ed. The Cognitive Neurosciences. London: MIT Press 1995; 463-474.
31. Tovée MJ. Is face processing special? Neuron 1998; submitted
32. Tanaka K, Saito H, Fukado Y, Moriya M. Coding the visual image of objects in the inferior temporal cortex of the macaque monkey. J Neurophysiol 1991; 66:170-189.

33. Fujita I, Tanaka K, Ito M, Cheng K. Columns for visual features of objects in monkey inferotemporal cortex. Nature 1992; 360 343-346.
34. Perrett DI, Oram MW. Neurophysiology of shape processing Image and Vision. Computing 1993; 11:317-333.
35. Wang G, Tanaka K, Tanifuji M. Optical imaging of functional organization in the monkey inferotemporal cortex. Science 1996; 272:1665-1667.
36. Stryker MP. Elements of visual perception. Nature 1992; 360:301-302.
37. Tanaka K. Inferotemporal cortex and higher visual functions. Curr Op Neurobiol 1992; 2:502-505.
38. Tovée MJ. An Introduction To The Visual System. Cambridge University Press 1996.
39. Rolls ET, Tovée MJ. The responses of single neurons in the temporal visual cortical areas of the macaque when more than one stimulus is present in the receptive field. Exp Brain Res 1995a; 103:409-420.
40. Sato T. Effects of attention and stimulus interaction on visual responses of inferior temporal neuron in the macaque. J Neurophysiol 1988; 60:344-364.
41. Sato T. Interactions of visual stimuli in the receptive fields of inferior temporal neurons in awake macaques. Exp Brain Res 1989; 77:23-30.
42. Gross GC, Rocha-Miranda CE, Bender DB. Visual properties of neurons in the inferotemporal cortex of the macaque. J Neurophysiol 1972; 35 96-111.
43. Perrett DI, Hietnan JK, Oram MW, Benton PJ. Organization and functions of cells responsive to faces in the temporal cortex. Phil Trans R Soc Lond 1992; 8 335:23-30.
44. Bruce CJ, Desimone R, Gross CG.Visual properties of neurons in a polysensory area in superior temporal sulcus of the macaque. J Neurophysiol 1981; 46:369-384.
45. Desimone R, Albright TD, Gross CG, Bruce C. Stimulus selective properties of inferior temporal neurons in the macaque. J Neurosci 1984; 4:2051-2062.
46. Perrett DI, Rolls ET, Caan W. Visual neurones responsive to faces in the monkey temporal cortex. Exp Brain Res 1982; 47:329-342.
47. Rolls ET, Baylis GC. Size and contrast have only small effects on the responses to faces of neurons in the cortex of the superior temporal sulcus of the monkey. Exp Brain Res 1986; 65:38-48.
48. Rolls ET, Tovée MJ, Purcell DG, Stewart AL, Azzopardi P. The responses of neurons in the temporal cortex of primates and face identification and detection. Exp Brain Res 1994; 101:473-484.
49. Yamane S, Kaji S, Kawano K. What facial features activate face neurons in the inferotemporal cortex of the monkey? Exp Brain Res 1988; 73:209-214.
50. Young MP, Yamane S. Sparse population coding of faces in the inferotemporal cortex. Science 1992; 256:1327-1331.

51. Rolls ET, Baylis GC, Leonard CM. Role of low and high spatial frequencies in the face-selective responses of neurons in the cortex in the superior temporal sulcus in the monkey. Vision Res 1985; 25:1021-1033.
52. Perrett DI, Smith PAJ, Potter DD, Mitelin AJ, Head AS, Mitner AD, Jeeves MA. Visual cells in the temporal cortex sensitive to face view and gaze direction. Proc R Soc Lond B 1985; 223:293-317.
53. Harries MH, Perrett DI. Visual processing of faces in the temporal cortex: Physiological evidence for a modular organization and possible anatomical correlates. J Cog Neurosci 1991; 3:9-24.
54. Perrett DI, Smith PAJ, Potter DD, Mistlin AJ, Head AS, Miiner AD, Jeeves MA. Neurones responsive to faces in the temporal cortex: Studies of functional organization sensitivity to identity and relation to perception. Human Neurobiology 1984; 3:197-208.
55. Desimone R, Gross CG. Visual areas in the temporal cortex of the macaque. Brain Res 1979; 178:363-380.
56. Baylis GC, Rolls ET, Leonard CM. Functional subdivisions of the temporal. J Neurosci 1987; 7:330-342.
57. Seltzer B, Pandya DN. Afferent cortical connections and architectonics of the superior temporal sulcus and surrounding cortex in the rhesus monkey. Brain Res 1978; 149:1-24.
58. Seltzer B, Pandya DN. Further observations on the parieto-temporal connections in the rhesus monkey Exp Brain Res 1984; 55:301-312.
59. Boussaoud DB, Ungerleider LG, Desimone R. Pathway for motion analysis: Cortical connections of the medial superior temporal and fundus of the superior temporal visual areas in the macaque. J Comp Neurol 1990; 296:462-495.
60. Perrett DI, Oram MW, Harries MH, Bevan R, Hictanen JK, Benson PJ.Viewer-centerd and object centerd coding of heads in the macaque temporal cortex. Exp Brain Res 1991; 86:159-173.
61. Heywood CA, Cowey A. The role of the "face-cell" in the discrimination and recognition of faces in monkeys. Phil Trans R Soc Lond B 1992; 8 335:31-38.
62. Campbell R, Heywood CA, Cowey A, Regard M, Landis T. Sensitivity to eye gaze with prosopagnosic patients and monkeys with superior temporal sulcus ablation. Neuropsychologia 1990; 28:1123-1142.
63. Cavada C, Goldman-Rakic PS. Posterior parietal cortex in rhesus monkey: 1. Parcellation of areas based on distinctive limbic and sensory corticocortical connections. J Comp Neurol 1989; 287:393-421.
64. Pandya DN, Yeterian FM. Architecture and connections of cortical association areas in the cerebral cortex. In: Peter A, Jones EG eds. Cerebral Cortex Vol 4 1985; New York: Plenum Press.
65. Johansson G. Visual perception of biological motion and a model for its analysis. Perception and Psychophysics 1973; 14:201-211.
66. Oram MW, Perrett DI. Responses of anterior superior temporal polysensory (STPa) neurons to "biological motion" stimuli. J Cog Neurosci 1994; 6:99-116.

67. Wachsmuth E Oram MW, Perrett DI. Recognition of objects and their component parts: Responses of single units in the temporal cortex of the macaque. Cereb Cortex 1994; 5:509-515.
68. Tanaka, K. Mechanisms of visual object recognition: Monkey and human studies. Curr Opin Neurobiol 1997; 7:523-529.
69. Desimone R. Face-selective cells in the temporal cortex of monkeys. J Cog Neurosci 1991; 3:1-8.
70. De Renzi E, Faglioni R, Groni D, Nicheiti P. Apperceptive and associative forms of protopagnosia. Cortex 1991; 27:213-221.
71. Tovée MJ. Face recognition: What are faces for? Current Biology 1995; 5:480-482.
72. Tovée MJ, Cohen-Tovée EM. The neural basis of prosopagnosia: A review. Cognitive Neuropsychology 1993; 10:505-528.

CHAPTER 3

Information Measures

Reading the Neural Code

It has long been assumed that different features of the environment are represented by different firing patterns of neurons in the brain. But it is unclear how information is encoded in the string of action potentials that stimulated neurons produce. Early studies of the activity of peripheral neurons seemed to demonstrate the existence of a neural code that represented the intensity of an external stimulus by the frequency of firing. These results have been developed into the rate coding or firing frequency theory, which implies that a neuron signals changes in the information it represents by changes in the number of impulses it produces in a period of time. The neuron integrates all the spikes reaching it over a certain sample period and produces a response over that period. Thus, each neuron functions as an "integrate-and-fire" device.[1] More recently it has been suggested that the temporal pattern of the spikes, and the interval between them, may also encode information about a stimulus.[2-7]

This has important implications. Not only could the amount of information encoded by a neuron about a particular stimulus feature be increased by utilizing additional information from a temporal code, but the temporal code could also signal information about another feature of the same object. For example, information about form and color could be brought together in neurons of V1, V2 and V4 by having neurons responding to, say, the form of a stimulus by firing rate and its color by the temporal patterns of its firing.[8] As previous studies of the segregation of color and form have been based on firing rate,[9,10] McClurkin and Optican argue that such an idea is not inconsistent with previous studies. McClurkin and Optican stress the close association of form and color at a comparatively early stage in the visual system and wax lyrical: "the red of rose is never perceived

The Speed of Thought: Information Processing in the Cerebral Cortex,
by Martin J. Tovée.

floating about unattached to the form of the petals". Whilst this statement may be true of normal human color vision, it is untrue in a variety of achromatopsic and other neurological conditions, which demonstrates that relatively circumscribed lesions can lead to a dissociation of color and form.[11] Moreover, in primates, in addition to physiological evidence suggesting a separation of color and form, lesion and anatomical studies have also shown a segregation of color and form within V1, V2 and V4.[12,15]

Pursuing the possibility of a temporal code, Barry Richmond and Lance Optican have applied principal component analysis and information theory to the analysis of spike trains from neurons in the primate lateral geniculate nucleus (LGN), striate cortex and temporal visual.[16-18] The general procedure was to smooth the spike train for a single trial with a Gaussian filter (with a σ of 10 msec initially, made adaptive in later papers; see Richmond and Optican[17]) to produce a spike density function. The spike density function was then sampled every 6 msec over 384 msec starting 20 msec after the onset of the visual stimuli, to produce a 64 point time series for each trial. At least five such time series were collected for each stimulus in the stimulus set. From these time series, the principal components were then extracted. The principal components were extracted from all individual responses of a neuron to all the test stimuli. The principal components form an orthogonal basis set and are ordered so that each component accounts for more variance than any subsequent one. The response of a neuron to a particular stimulus can then be fully described as a weighted sum of these principal components. The proportion of the variance accounted for by each principal component can be taken as a measure of its importance. In addition, to assess how many of the principal components are significant, a bootstrap randomization technique can be applied.[16]

The use of these techniques seemed to show that several principal components were required to describe the responses of a typical neuron to the different visual stimuli, which consisted of a set of black-and-white patterns based on Walsh functions (see Richmond, Optican, Podell, and Spitzer[19]). The first principal component was strongly correlated with the neuron's firing rate. The second and higher principal components displayed complex time courses, and Optican and Richmond[18] therefore suggested that information was contained not only in the magnitude of the firing rate response of a neuron, but also in the temporal pattern with which it fired. Optican and Richmond then analyzed the information encoded in this temporal variation and suggested that approximately 107% more information about a stimulus was available if the temporal pattern of firing of the neuron (as expressed by the first three principal components) to a stimulus over the whole 384 msec analysis period as compared to the information contained only in the firing rate.

Several problems suggest themselves when dealing with these studies. The first is a technical consideration. The comparatively small number of stimuli used in the study can result in an overestimation of the information encoded by neurons if a suitable correction factor is not used. Optican and Richmond in their initial study did not use a correction factor, but did in later papers. However, their correction factor might not have been appropriate. These problems were investigated in depth in a series of papers using neuronal responses from face-selective neurons in the temporal visual cortex.[20-22]

Implementing Principal Component Analysis

Data recorded from single neurons in the primate temporal visual cortex were initially represented as the poststimulus occurrence times of individual spikes, recorded with an accuracy of 1 msec. They were then smoothed by convolution with a Gaussian kernel with a σ of 5 msec (which corresponds to a low pass filtering with -3dB cutoff at 240 Hz). The smoothed data were then quantized, for some of the analyses, into bins of width 10 msec. As a result, the numbers of spikes in each bin *i*, giving the components x_i of a multidimensional response vector, were real rather than integers. In order to analyze the temporal course of the response, a certain number (referred to as B) of consecutive bins (usually B = 40) were selected to form the basis of the space of response vectors. The bins were chosen to cover either of two poststimulus time periods; 0-400 msec or 150-550 msec. Labeling the bins with the subscript i (e.g., i = 1,40), the response vector during a particular trial k is denoted as x_i^k. The covariance matrix of the responses recorded from a particular cell is:

$$C_{ij} = \sum_k (x_i^k - \bar{x}_i)(x_j^k - \bar{x}_j)/N$$

where N is the total number of trials for that cell, and x_i is the average number of spikes, over those N trials, falling in bin *i*. Eigenvalues and eigenvectors of the covariance matrix were extracted using a standard algorithm. The eigenvectors, labeled I = 1,..., B in order of decreasing eigenvalue, form a new basis for the response space.

Each subtracted response vector (i.e., with the average response vector subtracted out) was then expressed on this basis via multiplication by the appropriate orthogonal matrix. The resulting coefficients c_l^k (sometimes called the Karhunen-Loeve transform of the vector, $x_l^k - x^k$) are uncorrelated, in the sense that their covariance matrix is, by construction, diagonal. Note that although the eigenvectors of C_{ij}, the principal components, are normalized to unit length, they are plotted in the figures below after multiplication by the square root of the corresponding eigenvalue, in order to bring out their relative contribution to the variance.

The coefficients c_i^k were then averaged over the trials with the same stimulus. The means csi represent the loading of stimulus s on component i, and their statistical significance can be tested using the bootstrap method.[16] A scrambled stimulus-response pairing is obtained by assigning random selections of actually recorded responses to new stimulus labels, while maintaining the same number of trials of each stimulus as in the authentic pairing. The coefficients c_i^s are then calculated for each pseudo-stimulus label, with the number of labels high enough that the distribution of coefficients approaches, for each component, a continuous distribution (we used a hundred times as many labels for the bootstrap distribution as experimental stimuli). The statistical significance of the original can now be tested against the null hypothesis distribution that the relationship between stimulus and response be purely random. If there are 40 coefficients, then the response space is 40 dimensional. However, given that there were 20 stimulus conditions, if there were no noise, (i. e., the response were uniquely determined by the stimulus), there should be no more than 19 components that would be significant.

Calculating Information Measures

If S denotes the set of all stimuli s, and R the set of responses, the average information contained in the responses of a particular cell about that set of stimuli is defined as:

$$I(S,R) = \Sigma_{s \subset S} \Sigma_{r \subset R} P(s,r) \log_2 \frac{P(s,r)}{P(s)P(r)}$$

where P is the probability of occurrence of a particular event. In evaluating the information content from the data recorded, we consider different definitions of the response set. In one case, responses are simply quantified by the number of spikes within a preset time period (a unidimensional measure). In another case, consider a multidimensional response space spanning the first Q (typically $Q = 1, 2$, or 3) of the principal components extracted with the algorithm above. If 3 principal components are included, the particular response relative to trial k is defined as the triplet (C^k_1, C^k_2, C^k_3). Although the set of stimuli can be discrete (it is in the present experiment), R is generally a continuum (in the second case, a Q-dimensional vector) space. As in practice the expression for I has to be calculated by performing a sum rather than an integral, R needs to be quantized. This second quantitization follows a procedure similar to the first. Consider first the spike count case. The original data are represented by the number of spikes nk recorded in trial k within the prescribed window, minus the average over all trials. The range of the data is set as ($-\Delta n, \Delta n$), with Δn the largest between the maximum and absolute value of the minimum recorded for that cell. This range is divided into a preselected number D of bins (we ultimately used $D = 15$). D-2 bins have width $dn = 2\Delta n/(D-3)$, with the second bin centered around—Δn, the third shifted by dn, and so on. The first and last bins cover the two semiinfinite intervals at the extremes. Considering now only the trials relative to a given stimulus s, a smoothing procedure is applied by convolving the individual values n with a Gaussian kernel of width equaling the standard deviation $\sigma_n(s)$ of the values relative to the same stimulus. The result, normalized by dividing by the total number of trials, is quantized into the bins defined above, the area within each bin being used as an estimate of the joint probability P(s,r), where r corresponds to one of the response bins. Summing over all stimuli gives $P(r) = \sum_{s \subset S} P(s,r)$. In the case of the information contained in the *Q* principal components, the original data are represented by the Q dimensional set of coefficients c^k. In this case the average response has already been subtracted out. Repeating the same procedure Q times in order to find the ranges spanned by the coefficients for each component divides the ranges into bins, smooths the distribution of values recorded for each stimulus, and finds P(s,r). Each of the D^Q bins of the response space is now the product of independently determined bins for each of the Q components, and the relative probability is obtained by multiplying the estimated probabilities of the response falling into each unidimensional interval. As noted by Optican and Richmond,[18] it would be preferable to estimate P(s, r) by using the more time-consuming procedure of recalculating separate principal components for each stimulus, because these are in general different from those calculated over the whole stimulus set. However, this simplification in the construction of the estimator P(s,r) is of minor import when compared to the distortions produced by limited sampling.

Correction Factors

The fact that the probability distribution for each response is extracted from a small sample remains a problem, which cannot be solved by adjusting the kernel used for smoothing, nor by a judicious quantization into bins. This can be seen by using again a bootstrap procedure similar to the one used to test the significance of principal component coefficients.[23] One generates a scrambled stimulus-response pairing, with the number of pseudo-stimulus labels now equal to the number of real stimuli, and calculates the information I_0 contained in the responses about their randomly paired pseudo-stimuli. I_0 should be zero, whereas it turns out, in practice, to be of the same order as the original I. This is because the use of probability distributions based on a limited number of trials for each stimulus unavoidably biases an information estimate upwards, by producing random fluctuations in the distributions of responses to different stimuli, which are then picked up by the information measure as being related to the stimuli.[21]

The most straightforward remedy is to subtract I_0 from I, to obtain a measure $I_s = I - I_0$ which averages zero, even for small samples, in the absence of a stimulus-response causal relationship.

Optican et al[23] have argued that, instead of subtracting I_0, its square should be used (so $I_s' = I[I - (I_0/I)^2]$). Subtraction of the square (which is less than the fraction itself) leaves an estimate I_s' often closer to I than to I_S. Optican et al[23] argue in favor of subtracting the square on the basis of computer simulations (and data from two cells in V1), which seem to indicate a faster convergence to the asymptotic "correct" value. These computer simulations, based on different distributions than the unknown ones characterizing real data, could lead to inappropriate conclusions with small sample sizes (i.e., numbers of trials), although of course both correction procedures approach the same asymptotic correct value as the sample size increases.[21] Because the upward bias of I increases with the dimensionality of the response space, the use of some subtracted estimate is particularly important when calculating the information contained in the first Q principal components, with $Q > 1$.

To determine which correction procedure is the most appropriate, both correction procedures were applied to the same data set to allow comparison of the results (Fig. 3.1.). Use of the correction procedure $I_s = I - I_0$ (Correction 2), tended to underestimate the information when there are few (3-10) presentations of each stimulus, but was more likely to be accurate with 15-60 presentations of each stimulus. In contrast, the measure 1s' = $I[1 - (I_0/I)^2]$ (Correction 1), underestimated less than Correction 2 with small numbers of presentations of each stimulus (3-7), but did provide somewhat of an overestimate of the information about the stimulus for all other numbers of stimulus presentations in the range 10-60.

Principal Components

First, we consider the situation where the stimulus condition is one of four different faces, with one of five different fixation points (a total of twenty different stimulus conditions). The responses to these stimuli from 48 neurons in the macaque temporal visual cortex were recorded.[20] An example of the first four principal components from a single neuron are shown in Figure 3.3. The first principal component accounted for 20% of the variance and has a similar wave form to the spike density function (Fig. 3.2). This relationship of the firing rate to the first principal component can be further illustrated by plotting the firing rate in a 400 msec period against the weighting of the first two principal components (Fig. 3.4). Each point in this figure represents the firing rate versus the associated weight coefficient of the neuronal response to the stimulus for that principal component. There is a very strong correlation (0.99) between the first principal component and the firing rate. There was no correlation between the firing rate and the second principal component (a correlation factor of 0.11), or between the firing rate and any of the higher principal components. This relationship was true of all the other neurons in both this study[20] (see Fig. 3.5) and a subsequent study,[21] and was also consistent with the findings of Optican and Richmond.[18,23-24]

For the neuron shown in Figures 3.2 and 3.3, the second principal component accounted for 8.9% of the variance and had a biphasic component at around 100 msec, corresponding to the onset latency of the neuron to different stimulus conditions. After this initial biphasic part of the waveform, the remainder of the second component is close to zero, suggesting that it is not related to changes in

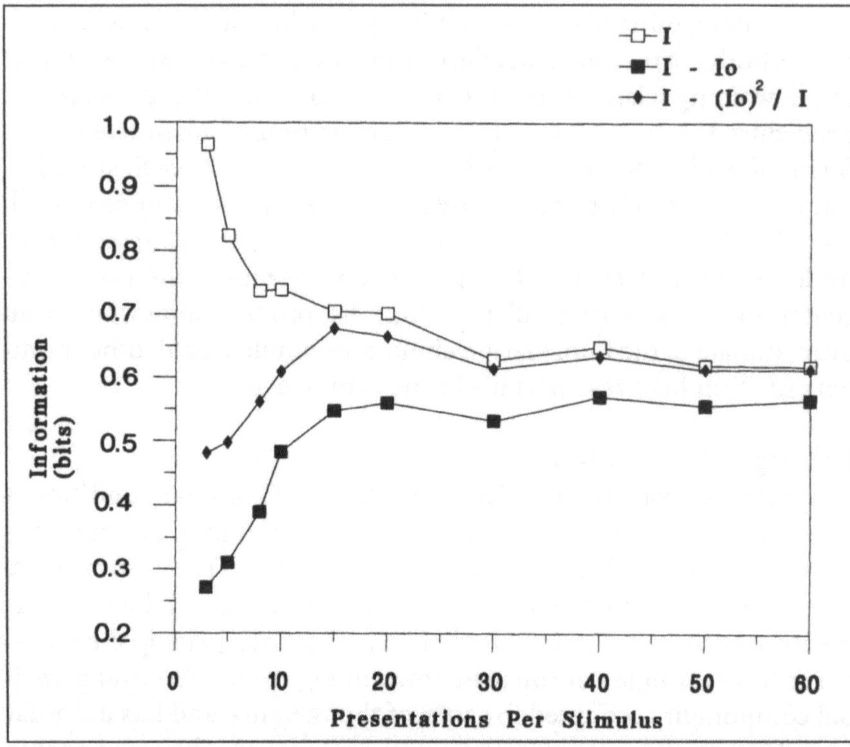

Fig. 3.1. A representation of the information on which stimulus was present (Information in bits, Is) based on the first three principal components plotted against the number of stimulus presentations, when there was no correction procedure (I), when correction procedure 1 was applied (I's = $I(1-(Io/I)^2)$), and when correction procedure 2 was applied (Is = I-Io). (Reproduced with permission from Tovée MJ, Rolls ET, Treves A, Bellis RP, J Neurophysiol 1993; 70:640-654. © (1993) The American Physiological Society).

the firing rate of the neuron after it has started to respond a stimulus. This waveform pattern for the second principal component was found for most of the neurons. Examination of the response latencies of neurons showed variations in latency with fixation position (and thus the position of the face stimulus on the retina). The response latencies with fixation position are shown for our example neuron in Figure 3.6. The neuron responded around 20 msec faster if the fixation point was at the center of the face rather than at the edge. The shape of the waveform of the second principal component closely resembled the difference in the latencies of the neural responses. The second principal component therefore does not seem to be related to the identity of the stimulus, but only to the temporal

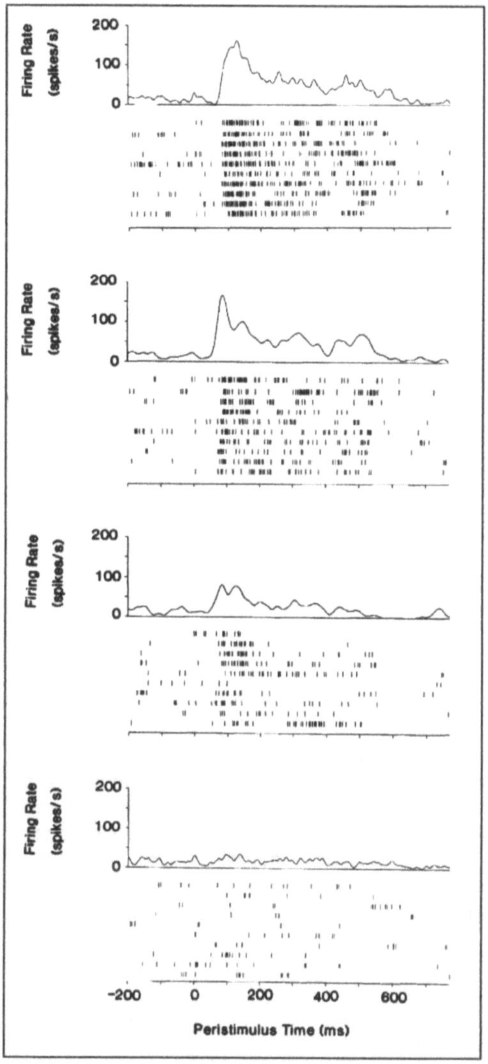

Fig. 3.2. Peristimulus rastergrams and spike-density histograms showing the response of a single neuron in the inferior temporal cortex to four different face stimuli. The face stimulus was presented at time 0 and was visible for 500 msec. Each horizontal raster shows the response of the neuron on a single trial presentation, and each vertical line represents 1 spike from the cell. (Reproduced with permission from Tovée MJ, Rolls ET, Treves A, Bellis RP, J Neurophysiol 1993; 70: 640-654. © (1993) The American Physiological Society).

aspects of the spike train, in as much as it reflected the response latencies of the neuron. This was confirmed by using a different stimulus set. Instead of using four different face stimuli and five different fixation positions, in this second set of experiments there were twenty different face stimuli and only one fixation position at the center of the image.[21] Under these stimulus conditions the waveform of the second principal component is flattened compared to htat from the first set of experiments (see Figs. 3.3 and 3.7), and it accounts for a reduced proportion of the variance and a reduced amount of information (see below).

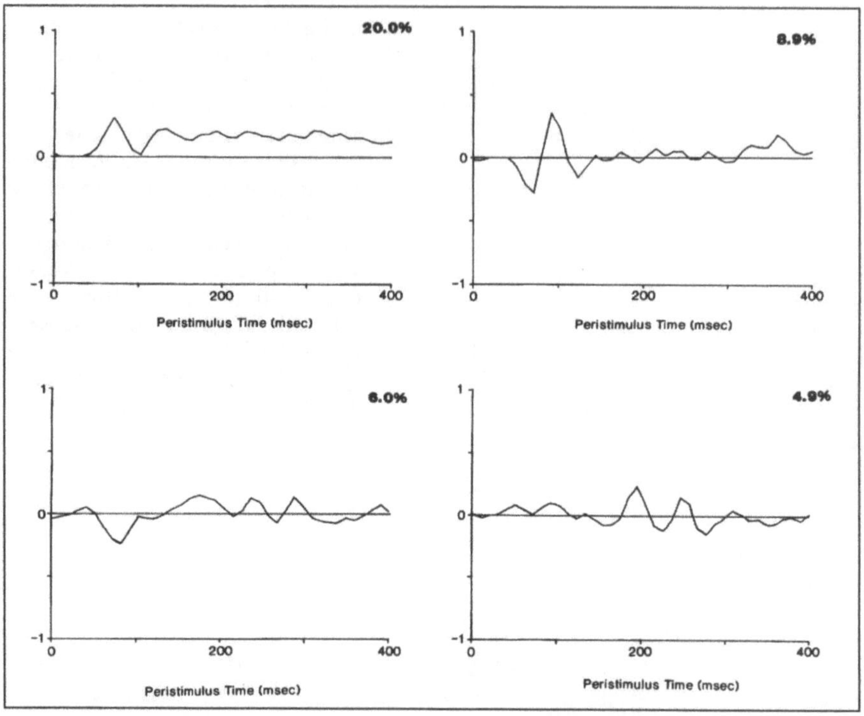

Fig. 3.3. The first four principal components extracted from the responses of the neuron shown in Figure 3.2 to four different visual stimuli fixated in five different positions (the center and all four corners of the picture). The percentage of the variance for each principal component is shown (principal component 1 is top left, principal component 2 is top right, principal component 3 is bottom left and principal component 4 is bottom right (Reproduced with permission from Tovée MJ, Rolls ET, Treves A, Bellis RP, J Neurophysiol 1993; 70: 640-654. © (1993) The American Physiological Society).

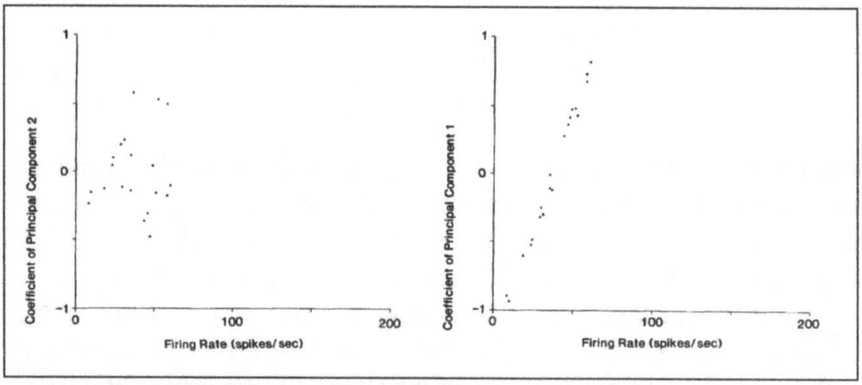

Fig. 3.4. For a single neuron, the relationship between the firing rate and the coefficients of principal components 1 and 2. (Reproduced with permission from Tovée MJ, Rolls ET, Treves A, Bellis RP, J Neurophysiol 1993; 70: 640-654. ©(1993) The American Physiological Society).

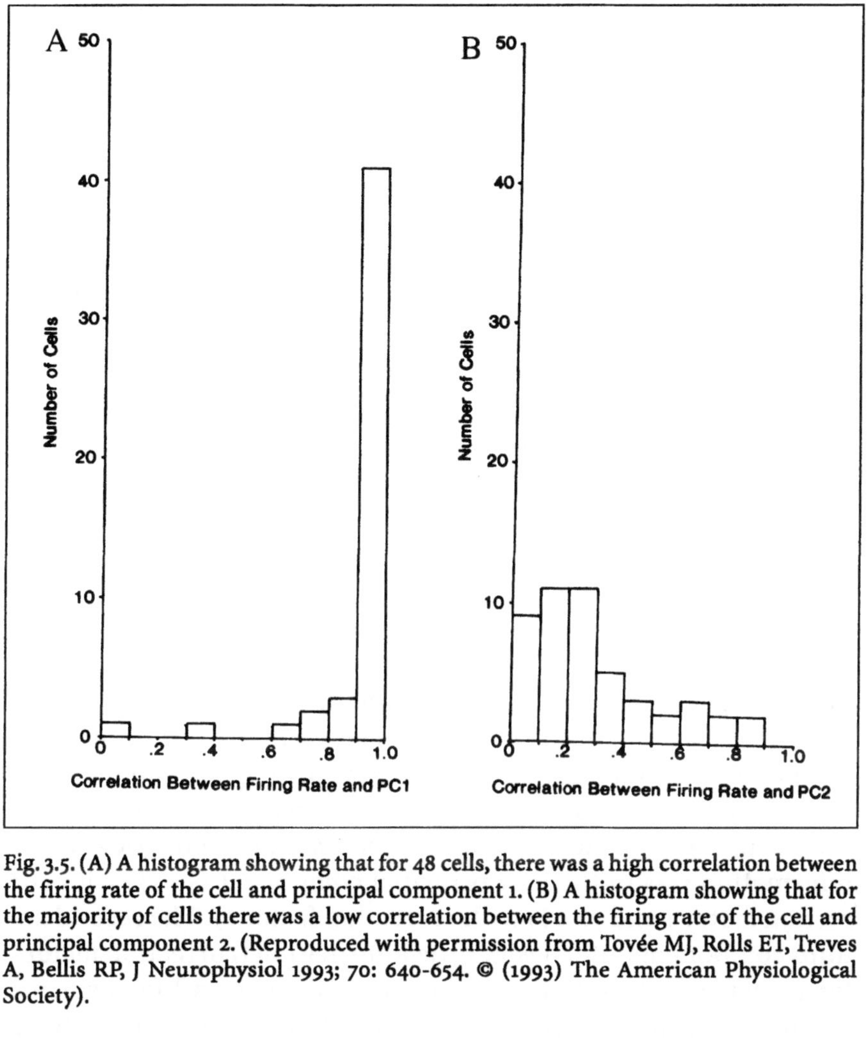

Fig. 3.5. (A) A histogram showing that for 48 cells, there was a high correlation between the firing rate of the cell and principal component 1. (B) A histogram showing that for the majority of cells there was a low correlation between the firing rate of the cell and principal component 2. (Reproduced with permission from Tovée MJ, Rolls ET, Treves A, Bellis RP, J Neurophysiol 1993; 70: 640-654. © (1993) The American Physiological Society).

For the cell illustrated in Figure 3.3, the third and fourth principal components account for only 6.0% and 4.9% of the variance, which are typical values for these lower principal components for all the cells tested. The weightings of the neuronal responses to a given stimulus on the principal components beyond the first three were not significantly above chance when assessed with the bootstrap procedure. Across the population of cells it was only the first principal component, and in some cases the second and third principal components, that were reliably linked to which stimulus was presented.

Information Measures and the Principal Components

The information measures were initially calculated over two 400 msec periods. The first period started from the onset of the stimulus (0-400 msec) and the second started from 150 msec after the onset of the stimulus (150-550 msec). Given that the neurons responded with a latency of 80-100 msec, the first time period will include the cell's response to the onset of the stimulus. The second time period will avoid the large change in firing rate engendered by the cell beginning to respond to the stimulus and will instead measure whether there is temporal information in the spike train once it has already begun firing.

For the first study, when there were four images and five fixation points, analysis of the principal components did seem to increase the total information available.[20] If no correction is applied to the data, then the second and third principal component can provide a significant proportion of the total available information in the 0-400 msec time sample (an increase of 62.5% for principal component 2 and 38.5% for principal component 3). However, if one uses either correction, then this is greatly reduced (correction factor 1: for principal component 2 an increase of 35.9% and for principal component 3 an increase of 13.5%; correction factor 2: for principal component 2 an increase of 18.8% and for principal component 3 an increase of 8.4%). This pattern is paralleled in the second study when the stimuli were 20 faces with a single fixation point. Most of the information was in the first principal component (56.1% with correction factor 1 and 72.2% with correction factor 2). The second and third components added very little information (correction factor 1: for principal component 2 an increase of 17.9% and for principal component 3 an increase of 16.1%; correction factor 2: for principal component 2 an increase of 8.4% and for principal component 3 an increase of 6.7%). As previously mentioned, the second principal component is strongly linked to response latency and, when the fixation position is constant and the difference in response latency between conditions is reduced, the proportion of information signaled by the second principal component is greatly reduced.

These findings suggest that temporal encoding, as measured by principal components, adds very little information to that available from firing rate. Based purely on firing rate, it is possible to "read off" the identity of a stimulus at the end of the ventral system.[25] An equivalent situation seems to exist in the dorsal system.

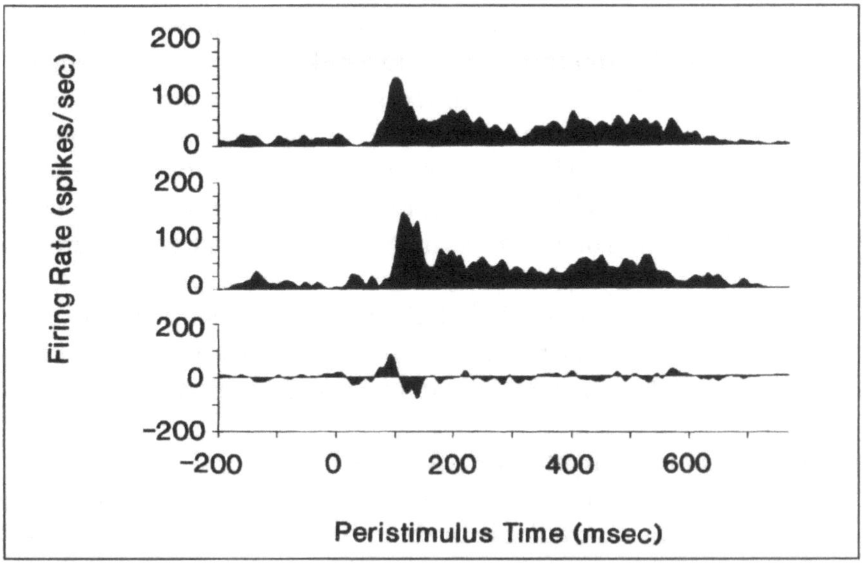

Fig. 3.6. Spike-density functions for a single neuron when the fixation point was at the center of the screen (top function), or at the edge of the image (middle function). The bottom function shows the difference between the two functions. The neuron's response latency seems to be around 20 msec longer when the fixation spot is at the edge of the stimulus (Reproduced with permission from Tovée MJ, Rolls ET, Treves A, Bellis RP, J Neurophysiol 1993; 70: 640-654.© (1993) The American Physiological Society).

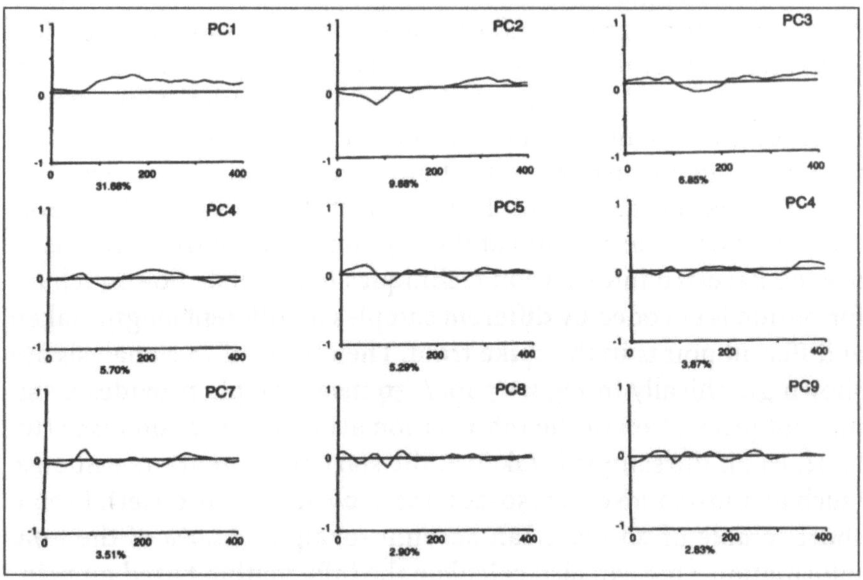

Fig. 3.7. The first 9 principal components extracted from the responses of one neuron to 20 different visual stimuli. The percentage of variance accounted for by each principal component is shown. The scale of the ordinate is constant and the abscissa shows peristimulus time (msec). (Reproduced with permission from Tovée MJ, Rolls ET. Visual Cognition 1995; 2:35-58 © (1995) Lawrence Erlbaum Associates Ltd).

Based purely on the firing rate of single neurons in V5 or MST, one can predict the behavior of the whole monkey on a motion discrimination task.[26-28] Furthermore, microstimulation of small groups of these neurons alters the performance of the monkeys on the discrimination task.[29-30] This microstimulation purely increases the firing rate of neurons with a specific response profile to bias the monkeys' representation of a visual stimulus. No clever pattern of stimulus is required; it is enough merely to increase the average firing rate over a particular time period. This does raise questions as to the importance, or even existence, of a temporal code as envisioned by researchers such as Lance Optican or Barry Richmond.

How Fast Is Information Processed?

A significant amount of the total information available in a neuron's response is available in short time periods (20 to 50 msec) at the start of the neuron's response. This can be illustrated in two ways. Firstly, one can determine the correlation between the firing rate in a short time period and the mean firing rate over a 500 msec sample (100-500 msec poststimulus presentation). As can be seen in Figure 3.8, in the case of this example neuron (the same cell as shown in the previous figures) there is a strong correlation between samples of 100 msec and 50 msec and the mean rate in the 500 msec period (0.95 and 0.84 respectively). The correlation was good even for a 20 msec sample (0.72). The correlation was also calculated for a population of 48 cells and is shown in Figure 3.9. For most of the cells there was a good correlation for both 50 and 20 msec samples.

The second and more direct way of determining the importance of short time samples at the beginning of the spike train is to use the Bayesian information techniques to calculate how much information is encoded by different samples of different lengths taken at different points in the spike train. The results of this analysis are shown graphically in Figure 3.10. A 50 msec sample provides a significant proportion of the information available over 400 msec (up to 84.4%) if the sample is taken at the start of the neuron's response (such as 100-150 msec or 150-200 msec poststimulus onset). Even a short sample of 20 msec can account for up to 64.9% of the total information. One can also calculate the information based on principal components over different time samples. Such an analysis is shown if Figure 3.11. The information was calculated based on the first three principal components. The results showed the same pat-

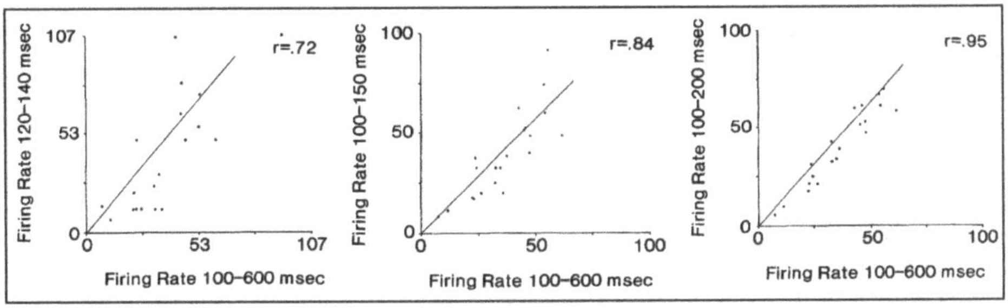

Fig. 3.8. The relationship between the firing rates of a neuron measured in short samples (20, 50 and 100 msec) with the mean firing rate measured in a 500 msec period. The correlation coefficient (r) is shown in each case. (Reproduced with permission from Tovée MJ, Rolls ET, Treves A, Bellis RP, J Neurophysiol 1993; 70: 640-654. © (1993) The American Physiological Society).

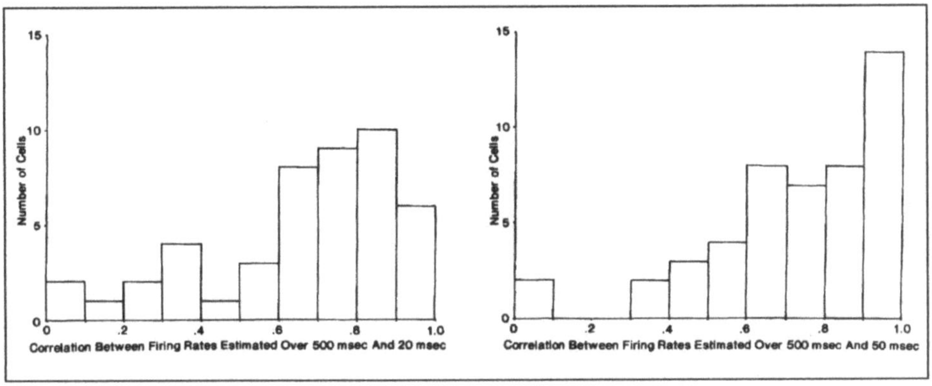

Fig. 3.9. Histograms illustrating for the population of 48 cells analyzed the correlation between the firing rates estimated over 20 msec (left) and 50 msec (right) and 500 msec. (Reproduced with permission from Tovée MJ, Rolls ET, Treves A, Bellis RP, J Neurophysiol 1993; 70: 640-654. © (1993) The American Physiological Society).

tern as found in information from spike rates. For example, the information calculated from the from a 50 msec sample accounted for 77.1% of that available from a 400 msec sample.

These results all suggest that most of the information available from the analysis of a spike train is available in a short period at the beginning of the response. This can be further illustrated by plotting the cumulative information signaled by the spike train in 50 msec samples (Fig. 3.12). The information increases rapidly at first, but soon levels out to a plateau.

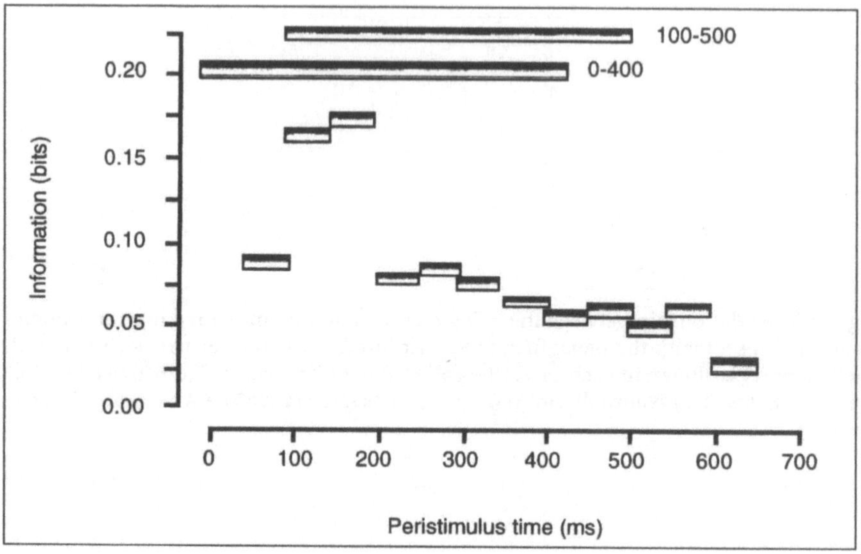

Fig. 3.10. The information available (using correction factor 1) from different 50 msec and 400 msec samples of neuronal activity. The length of the sample and the time at which it was taken with respect to the stimulus onset is shown by the length and position of each horizontal line. (Reproduced with permission from Tovée MJ, Rolls ET, Treves A, Bellis RP, J Neurophysiol 1993; 70: 640-654. © (1993) The American Physiological Society).

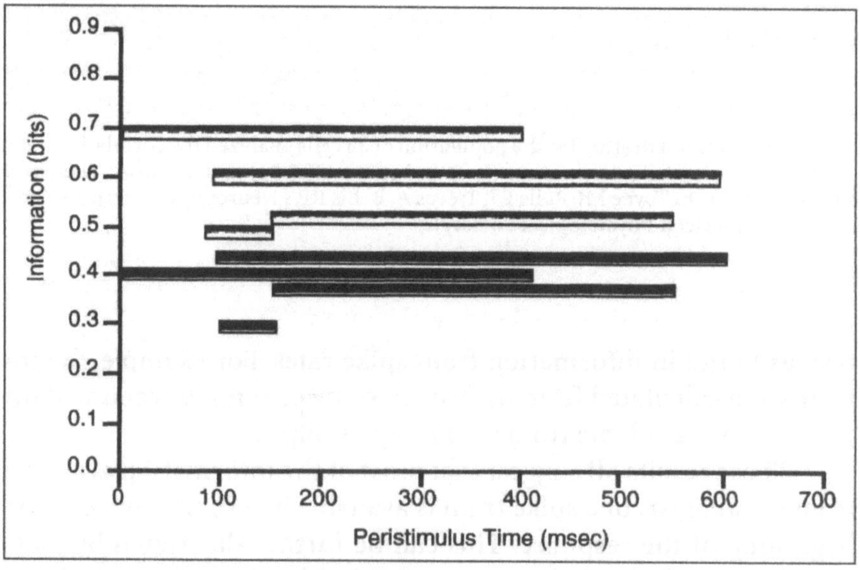

Fig. 3.11. The information available (using correction factor 1) about which image was shown and where it was fixated from the first three principal components (pale lines) and from the firing rate (dark lines) for different poststimulus samples. The data are the means from 48 cells. (Redrawn from Tovée MJ, Rolls ET, Treves A, Bellis RP, J Neurophysiol 1993; 70: 640-654.

Are Principal Components a Valid Way of Measuring Temporal Encoding?

Examination of the principal component analysis does suggest that, with suitable correction factors for the small sample size, only the first few principal components, possibly only the first principal component, reliably signal a significant amount of information about a stimulus. As the first principal component is based on firing rate, it seems that very little information is actually signaled by a temporal pattern. A further problem with the use of principal components is that they may over-estimate the amount of temporal information available.

Principal component analysis has been used to study the waveforms of evoked potentials as a method of extracting independently varying components from time series data. However, this use has been criticized because the information encoded in one component may also be extracted by another component which overlaps it in time, once the waveform has undergone principal component analysis and been decoded.[31] This problem also seems to apply to the use of principal component analysis to extract temporal patterns from spike trains. Additionally, Fotheringhame and Young[24] have suggested that each principal component can act as a "noised filter". The first principal component will capture variance which is primarily due to signal but will also represent some noise variance. Thus, the signal is combined with a certain temporal pattern of noise. This process continues through a series of filters, each one viewing the signal through a different filter of noise. Messages produced by different stimuli may be only poorly discriminated when combined with one set of obscuring noise, but when combined with another independent noise pattern become more discriminable. Because the noise represented by each PC is independent, the extra information conveyed by higher principal components is not redundant, but carries additional information to that carried by the first principal component.

Is Temporal Encoding Physiologically Possible?

The most extreme form of temporal coding, otherwise known as coincidence detection or interval coding, is characterized by information being carried in the precise timing of the occurrence of each action potential. This would radically increase the amount of information carried by a spike train.[32] A recent study by Softky and

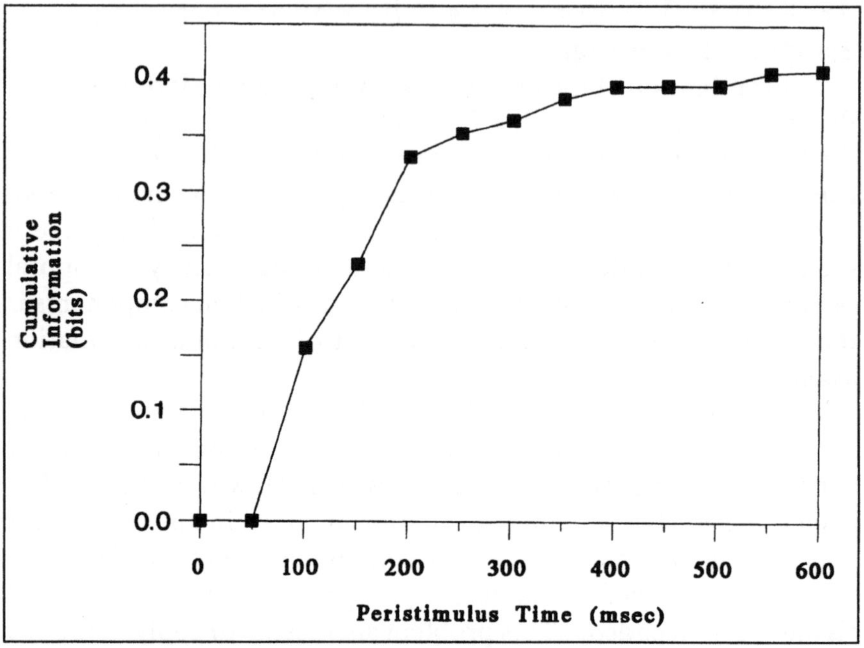

Fig. 3.12. Cumulative information available (using correction factor 1) from the neuronal responses taken in 50 msec samples. The data is an average from 48 cells. (Reproduced with permission from Tovée MJ, Rolls ET, Treves A, Bellis RP, J Neurophysiol 1993; 70: 640-654. © (1993) The American Physiological Society).

Koch[33] has been interpreted as evidence for a temporal code. They examined spike trains from primate V1, using a normalization procedure which allowed them to estimate the variability of the interspike interval (ISI) at nominally constant firing rates. They found that the spiking patterns approximated to a random (Poisson) process. They suggest that such irregularity is not possible with an integrate-and-fire mechanism, as a neuron which integrates random presynaptic events to reach spike threshold cannot preserve such irregularity in its own spike output. Therefore, cortical neurons must perform some sort of coincidence detection, such that a particular combination of presynaptic events leads to a postsynaptic spike. This combination would occur irregularly, so leading to the observed irregular ISI.[32-33] Such a neuron would be capable of transmitting information in the precise timing of individual spikes or their temporal pattern. However, although Softky and Koch find a Poisson distribution at constant firing rates, for an analysis of a 20 msec period when a cell is firing maximally to a stimulus, the distribution of spikes is not a Poisson distribution.[34] Moreover, it is possible that a balance of excitation and inhibition will also yield an irregular ISI from a standard integrate-and-fire neuron.[1,35] This can be explained in terms of random walk or diffusion process models of integrate-and-fire neurons.[36-37] It is suggested that the membrane potential undergoes a "random walk" between the resting potential and the spike threshold. The excitatory postsynaptic potentials (EPSPs) drive the potential toward spike threshold and the inhibitory postsynaptic potentials (IPSPs) drive it away from threshold. Depending upon the balance of excitation and inhibition, the ISI can potentially be very irregular.

The coincidence detector theory requires the effective set of synaptic events to occur at roughly the same frequency as the postsynaptic neuron's spike rate. Noncoincident EPSPs must either arrive infrequently or be prevented from summating to spike threshold by a very short membrane time constant (less than a few msec).

For a temporal code to function, a presynaptic signal must be preserved in the postsynaptic neuron. An obvious limiting factor here will be the electronic properties of the postsynaptic membrane.[1,38] Rapid changes in current will be dependent on the membrane time constant. The effect of a high time constant will be to apply a low-band pass filter on the temporal code. A passive cable model of the electrotonic properties of a neuron would suggest that the time constant will be lowest at the cell body and will rise with distance from it. This would suggest that synaptic connections to distal dendrites would not be able to support a temporal code. However, it has been shown that these synapses can alter the somatic membrane potential, presumably by active dendritic conductances.[39-41]

As to whether the basic time constant of a membrane, whether of the dendrite or cell body membrane, actually acts as a limit to temporal coding, it is hard to say. A wide range of time constants have been reported and the situation is further complicated by evidence that background synaptic activity significantly alters the time constant by a factor of 10.[42] Several authors have suggested that the membrane constant is reduced to a point which allows coincidence detection when neurons are active within their normal set of connections.[5,43] Additionally, as Fotheringhame and Young[38] point out, there is a suggestive relationship between the reported time constants of neurons and the function of the cell. Cells specialized for dealing with high frequency inputs, such as the hair cell of the inner ear, have a membrane time constant in the range 2.5 to 3.1 msec.[44] This can be compared with values for cerebellar cells of 50 msec, cortical pyramidal cells of 12.0 to 19.5 msec, and hippocampal cells of 28 to 66 msec.[45-47]

The coincidence detection theory also faces two other problems. The first is the noisy nature of neural activity and the lack of reliability in synaptic transmission. As stated above, for a temporal code to function there must be a faithful reproduction of the spikes across a synapse. However, far less than 100% of the spikes arriving at a synapse produce a postsynaptic spike, as roughly 10-40 excitatory responses are required to take a cortical neuron from its resting potential to its firing threshold.[48-50] Individual cortical neurons have from between 4,000 and 10,00 connections depending upon cortical area and species, of which up to 85% are excitatory.[51-52] The cell's membrane potential at any one time will be dependent on the activity across all its inputs, and if only a small proportion of these excitatory connections are active at any one time, this still means possibly hundreds of EPSPs occurring at the neuron in a very short time period.[1] This would contaminate with extraneous inputs the set of coincident events carrying information. Even if all these neurons are largely inactive, save for the single neuron providing the input using a temporal code, the spontaneous discharges of the other neurons will provide an unstable baseline from which to detect the postsynaptic potentials forming the temporal code.

The nature of temporal coding may be in the eye of the beholder. In firing rate coding, a stimulus is represented by a change in firing rate, but this is of course relative to the firing rate when the stimulus is not present. So if we use the term loosely, a temporal code could be based not on single spikes as the basic signaling unit, but upon the summed activity in short integration windows. In such a scheme, the ISI is random and encodes no information about a stimulus. The signal is encoded by differences in the level of activity in different integration windows. A longer integration window means that the code will be more accurate, as the effects of transmission noise will be minimized. However, there must be an upper limit to the window for biophysical, encoding and functional reasons. If the time window for integrating spikes extends beyond the membrane time constant, the postsynaptic cell will need to develop some form of "memory" mechanism to store the summed input. Moreover, during the response of a neuron there are changes in firing rate, and a long time window will lose these finer nuances in firing rate. Functionally, one needs comparatively short integration windows for the nervous system to be able to respond to changes in the environment and to able to control movement and motor function.

It seems likely that the nervous system compromises between a coincidence detection code and a firing rate code, by using a temporal code that is not based on the temporal spacing between action potentials, but on the difference in mean firing rate between different integration windows. The length of these windows will be very short, in the order of milliseconds, rather than hundreds of milliseconds. Indeed, it is likely that the integration window will coincide with the membrane time constant, as this will avoid the necessity of a specific new storage mechanism. As different neurons seem to have different membrane time constants, this will mean that different cells in different sensory or cognitive processing systems will have different integration windows which will be "customized" to their function. As discussed above, the membrane time constant changes with activity in the system and therefore so will the integration windows, possibly allowing the length of the window to be cut to fit the dynamic requirements of a functioning system. This flexibility would optimize the balance between temporal resolution and the intrinsic noise in neural communication: longer windows for low activity situations when the signal-to-noise ratio will be poor and short windows for higher activity situations when the signal-to-noise ratio will be better.

References

1. Shadlen MN, Newsome WT. Noise neural codes and cortical organization. Curr Op Neurobiol 1994; 4:569-579.
2. McClurkin JW, Optican L, Richmond BJ, Gawne T. Concurrent processing and complexity of temporally encoded neuronal messages in visual perception. Science 1991; 25:675-677.
3. Bialek W, Rieke F. Reliability and information transmission in spiking neurons. Trends in Neurosciences 1992; 15:428-434.
4. Richmond BJ, Gawne TJ, Jin G-X. Neuronal codes: Reading them and learning how their structure influences network organization. Biosystems 1997; 40:149-157.
5. Sejnowski TJ. Anatomical specializations of neurons and their contribution to coincidence detection. In: Konnerth A, Tsien RY, Mikoshiba K, Altman J eds. Coincidence Detection in the Nervous System. Strasbourg: HFSP 1996; 151-159.

6. Softky WR. Simple codes versus efficient codes. Curr Op Neurobiol 1995; 5:239-247.
7. van Steveninck RRD, Lewen GD, Strong SP, Koberle R, Bialek W. Reproducibility and variability in neural spike trains. Science, 1997; 275:1805-1808.
8. McClurkin JW, Optican LM. Primate striate and prestriate cortical neurons during discrimination 1. Simultaneous temporal encoding of information about color and pattern. J Neurophysiol 1996; 75:481-495.
9. Zeki S. Color coding in the cerebral cortex: The responses of wavelength-selective and color-coded cells in monkey visual cortex to changes in wavelength composition. Neuroscience 1983; 9:741-765.
10. Gallant JL, Braun J, Van Essen DC. Selectivity for polar hyperbolic and cartesian gratings in macaque visual cortex. Science 1993; 259:100-103.
11. Zeki S. A Vision of the Brain. Oxford: Blackwell Scientific Publications. 1993.
12. Shipp S, Zeki S. Segregation of pathways leading from leading from V2 to V4 and V5 of macaque monkey visual cortex. Nature 1985; 315:322-325.
13. Zeki S, Shipp S. The functional logic of cortical connections. Nature 1988; 335:311-317.
14. Schiller PH, Lee K. The role of primate extrastriate V4 in vision. Science 1991; 251:1251-1253.
15. Ghose GM, Ts'o DY. Form processing modules in primate area V4. J Neurophysiol 1997; 77:2191-2196.
16. Richmond BJ, Optican LM. Temporal encoding of two-dimensional patterns by single units in primate inferior temporal cortex II. Quantification of response waveform. J Neurophysiol 1987; 57:147-161.
17. Richmond BJ, Optican LM. Temporal encoding of two-dimensional patterns by single units in primate inferior temporal cortex II. Information transmission. J Neurophysiol 1990; 64:370-380.
18. Optican LM, Richmond BJ. Temporal encoding of two-dimensional patterns by single units in primate inferior temporal cortex. III. Information theoretic analysis. J Neurophysiol 1987; 57:162-178.
19. Richmond BJ, Optican LM, Podell M, Spitzer H. Temporal encoding of two-dimensional patterns by single units in primate inferior temporal cortex. I. Response characteristics. J Neurophysiol 1987; 57:132-146.
20. Tovée MJ, Rolls ET, Treves A, Bellis RP. Information encoding and the responses of single neurons in the primate temporal visual cortex. J Neurophysiol 1993; 70:640-654.
21. Tovée MJ, Rolls ET. Information encoded in short firing rate epochs by single neurons in the primate temporal visual cortex. Visual Cognition 1995; 2:35-58.
22. Panzeri S, Biella G, Rolls ET, Skaggs WE, Treves A. Speed, noise, information and the graded nature of neuronal responses. Network-Computation in Neural Systems 1996; 7:365-370.

23. Optican LM, Gawne TJ, Richmond BJ, Joseph PJ. Unbiased measures of transmitted information and channel capacity from multivariate neuronal data. Biol Cybern 1991; 65:305-310.
24. Fotheringhame D, Young MP. Is temporally coded information in spike trains due to the signal or the noise? J Comput Neurosci 1998: in press.
25. Rolls ET, Treves A, Tovée MJ. The representational capacity of the distributed encoding of information provided by populations of neurons in the primate temporal visual cortex. Exp Brain Res 1997; 114:149-162.
26. Britten KH, Shadlen MN, Newsome WT, Movshon JA. The analysis of visual motion. J Neurosci 1992:12:4745-4765
27. Celebrini S, Newsome WT. Neuronal and psychophysical sensitivity to motion signals in extrastriate area MST of the macaque monkey. J Neurosci 1994; 14:4109-4124.
28. Shadlen MN, Newsome WT. Motion Perception—Seeing and deciding. Proc Natl Acad Sci USA 1996; 93:628-633.
29. Salzman CD, Newsome WT. Neural mechanisms for forming a perceptual decision. Science 1994; 264:231-237.
30. Celebrini S, Newsome WT. Microstimulation of extrastriate area MST influences performance on a direction discrimination task. J Neurophysiol 1995; 73:437-448.
31. Wood CC, McCarthy G. Principal component analysis of event-related potentials: Simulation studies demonstrate misallocation of variance across components. Electroenceph Clin Neurophysiol 1984; 59:249-260.
32. Softky WR. Sub-millisecond coincidence detection in active dendrite trees. Neuroscience 1994; 58:13-41.
33. Softky WR, Koch C. The highly irregular firing of cortical cells is inconsistent with temporal integration of random EPSPs. J Neurosci 1993; 13:334-350.
34. Rolls ET, Treves A, Tovée MJ. The representational capacity of the distributed encoding of information provided by populations of neurons in the primate temporal visual cortex. Exp Brain Res 1997; 114:149-162.
35. Shadlen MN, Newsome WT. Is there a signal in the noise? Curr Op Neurobiol 1995; 5:248-250.
36. Gerstein G, Mandelbrot B. Random walk models for the spike activity of a single neuron. Biophys J 1964; 4:41-68.
37. Calvin W,Stevens C. Synaptic noise and other sources of randomness in motoneuron interspike intervals. J Neurophysiol 1968; 31:574-587.
38. Fotheringhame DK, Young MP. Neural coding schemes for sensory representation: Theoretical proposals and empirical evidence. In: Rugg M, ed. Cognitive Neuroscience. Psychology Press 1997; 47-76.
39. Kim H, Connors B. Apical dendrites of the neocortex: Correlation between sodium and calcium dependent spiking and pyramidal cell morphology. J Neurosci 1993; 13:5301-5311.

40. Amitai Y, Friedman A, Connors B, Gutnick M. Regenerative activity in apical dendrites of pyramidal cells in the neocortex. Cereb Cortex 1993; 3:26-38.
41. Stuart G, Sakman B. Active propagation of somatic action potentials into neocortical pyramidal cell dendrites. Nature 1994; 367:69-72.
42. Bernander O, Douglas RJ, Martin KAC, Koch C. Synaptic background activity influences spatiotemporal integration in single pyramidal cells. Proc Natl Acad Sci USA 1991; 88:11569-11573.
43. Nowak LG, Sanchez-Vives MV, McCormick DA. Influence of low and high frequency inputs on spike timing in visual cortical neurons. Cereb Cortex 1997; 7:487-501.
44. Preyer S, Hemmert W, Pfister M, Zenner HP, Gummer AW. Frequency response of mature guinea-pig outer hair cells to stereociliary displacement. Hearing Res 1994; 77:116-124.
45. Rapp M, Segev I, Yarom Y. Physiology morphology and passive models of guinea-pig cerebellar Purkinje cells. J Physiol (Lond)1994; 474:101-118.
46. Mason A, Larkman A. Correlations between morphology and electrophysiology of pyramidal neurons in slices of rat visual cortex II. Electrophysiology. J Neurosci 1990; 10:1415-1428.
47. Spruston N, Johnston D. Perforated patch clamp analysis of the passive membrane properties of three classes of hippocampal neurons. J Neurophysiol 1992; 67:508-529.
48. Connors B, Gutnick M, Prince D. Electrophysiological properties of neocortical neurons in vitro. J Neurophysiol 1982; 48:1302-1320.
49. Mason A, Nicoll A, Stratford K. Synaptic transmission between individual pyramidal neurons of the rat visual cortex in vitro. J Neurosci 1991; 11:72-84.
50. Stevens CF. Cooperativity of unreliable neurons. Current Biology 1994; 4:268-269.
51. Peters A. Number of neurons and synapses in primary visual cortex. In: Jones A, Peters A eds. Cerebral Cortex, Further Aspects of Cortical Function Including Hippocampus. New York: Plenum 1987; 267-294.
52. Beaulieu C, Kisvardy Z, Somogyi P, Cynader M, Cowey A. Quantitative distribution of GABA-immunopositive and immunonegative neurons and synapses in the monkey striate cortex (V1). Cereb Cortex 1992; 2:295-309.

CHAPTER 4

Visual Masking

Limiting the Time for Visual Processing

The phenomenon of visual masking is one well known in human psychophysics. Masking can occur in two ways: backward or forward masking. In backward masking, a test stimulus is briefly presented followed at a variable interval by a masking stimulus. As the interval between the onset of the test stimulus and the onset of the mask stimulus (a period of time called the stimulus onset asynchrony or SOA) is reduced, the difficulty of correctly recognizing the test stimulus and details of its parameters is increased. In forward masking, the mask stimulus precedes the test stimulus. Reducing the interval between the offset of the mask and the onset of the test stimulus reduces the saliency of the test image. These psychophysical paradigms have been used to investigate the temporal aspects of visual processing, including face perception. These latter studies have shown not only how rapidly we can recognize faces, but also suggests those parameters of the stimulus which alter the speed of recognition. For example, under conditions of backward masking it is harder to detect the presence of an inverted face than an upright face, and it is harder to recognize a face whose features have been rearranged than a normal face.[1-2] This effect is called the face-detection effect (FDE), and illustrates the importance of the orientation and feature configuration in face processing. A similar phenomenon is the face-classification effect (FCE) in which subjects are asked to classify whether a face is normal or rearranged under masking conditions. It is easier to classify normal faces than rearranged faces.

The combination of masking paradigms with electrophysiology allows a number of useful lines of investigation to be explored. If, as believed, the masking stimulus does interfere with the neural processing of the test stimulus, then the masking paradigm may be used to explore the speed of processing of visual information at

The Speed of Thought: Information Processing in the Cerebral Cortex, by Martin J. Tovée.

individual synapses. Merely to make brief presentations of a stimulus is a poor method of deducing processing time, as the activity of visual neurons is not limited by the brevity of the stimulus presentation. For example, experiments show that face-selective neurons in the primate temporal cortex will respond for several hundred milliseconds to a 16 msec presentation of a face stimulus (see below). This finding is consistent with the results of psychophysical studies which have suggested that the neuronal representation of a briefly presented pattern (less than 6 msec) outlasts its physical presence by anything from 400 to 1,500 msec.[3] To measure the minimum amount of time required to process a visual stimulus one must use a task where the time available to process the stimulus is limited. A backward visual masking paradigm will allow the time available for processing to be controlled, so the minimum time necessary to mediate recognition can be determined.

Masking can also be used to investigate the role of feature configuration in object processing, both at the single cell and population level. However, very little interest has been expressed in using the combination of masking paradigms with electrophysiology until very recently, when they have been utilized to study higher level processing in the primate visual system.[4-6] This chapter will describe in detail these experiments and the implications they have for higher-level visual processing. These experiments fall into three distinct phases: firstly, a series of psychophysical experiments using human subjects to determine the behavioral responses of subjects in the backward masking paradigm: next, primate electrophysiology using the same paradigm and stimulus delivery system to determine the effect of backward and forward masking on the responses of individual neurons; and lastly, the differences in a neuron's response to normal and rearranged stimuli and how masking affects these differences.

A Psychophysical Study of Backward Visual Masking

In this experiment, eight human observers made two forced-choice judgements.[4] The first judgment was whether the face stimulus was a normal or a rearranged face, and the second judgment was to identify whose face they thought had been presented. It had previously been reported that a significant proportion of face-selective neurons were responsive to facial configuration.[7-10] Thus, the responses of the subjects can be directly compared to the responses of the face-selective neurons.

The stimuli used were static visual stimuli subtending 8.0° in the visual field presented on a video monitor at a distance of 1.0 m. The fixation spot position was at the center of the screen, and the monitor was viewed binocularly, with the whole screen visible to both eyes. The masks used in these experiments were either a face stimulus or a pattern stimulus consisting of overlapping N and O letters (see Fig. 4.1). This masking pattern was used because it is similar to the mask used in the previous psychophysical experiments investigating the FDE and FCE effects. Five different faces were used as stimuli (four men and one women). All the faces were well known to each of the eight observers used in the experiment. The stimuli were either normal or rearranged faces. At all times, the mean luminance of whatever was displayed on the monitor (even when it was a blank screen) was set to the same level, so that there were no luminance changes during each trial. This was important, as differences in luminance levels between the test and mask stimulus can alter the effectiveness of the masking paradigm.

Given the importance of the temporal relationship between the test and the mask stimuli, the test paradigm will be described in detail. Each trial started 500 msec prior to the onset of the test image, with a 500 msec warning tone to allow fixation of the fixation point, which appeared simultaneously (see Fig. 4.2). At 100 msec prior to the onset of the test image, the fixation spot was blinked off. As a result, there was no stimulus on the screen in the 100 msec period immediately preceding the test image which could interfere with its perception. The mean luminance of the monitor's screen was kept constant, so that the test and mask stimuli had the same luminance levels. The tone was then switched off and the test image was switched on for 16 msec. This 16-msec period was the frame interval of the video framestore from which the images were presented. The image was drawn on the monitor from the top to the bottom in the first 16 msec of the frame period by the framestore, with the remaining 4 msec of the frame period being the vertical blank interval. The monitor had a persistence of less than 3 msec, so that no part of the test image was present at the start of the next frame. SOA values of 20, 40, 60, 100 or 1000 msec (chosen in a random sequence by the computer) were used. The duration of the masking stimulus was 300 msec. Three hundred stimulus presentations (trials) were made, randomized in blocks of 50 presentations.

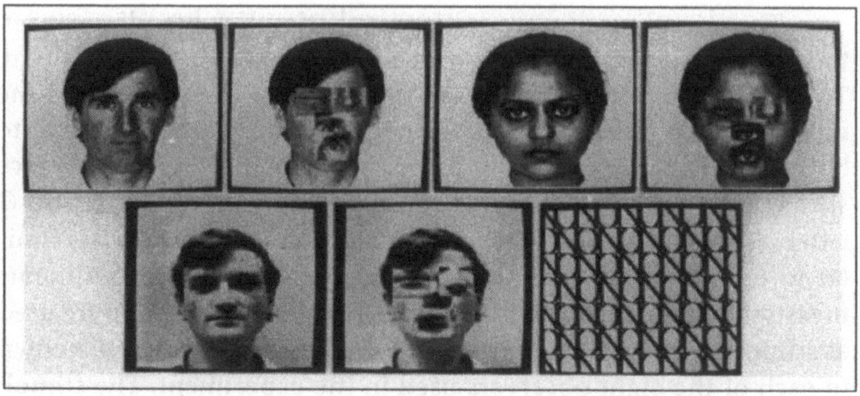

Fig. 4.1. Examples of the test images used in the normal spatial configuration and with rearranged features. The N-O pattern mask is also shown. (Reproduced with permission from Rolls ET, Tovée MJ, Purcell DG, Stewart AL, Azzopardi P. Exp Brain Res 1994; 101:473-484. © (1994) Springer-Verlag).

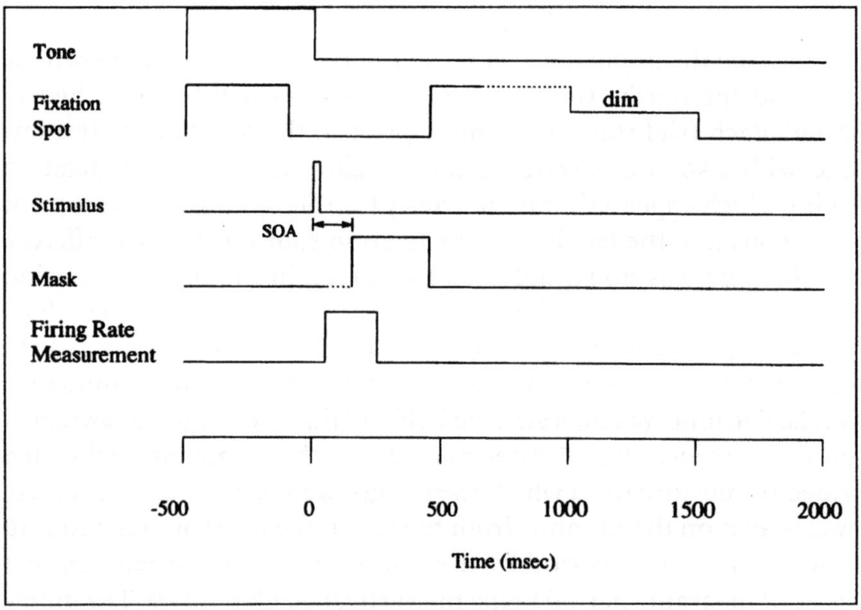

Fig. 4.2. The timing used in the backward masking visual fixation blink task. (Reproduced with permission from Rolls ET, Tovée MJ, Proc R Soc Lond B 1994; 257: 9-15. © (1994) Royal Society).

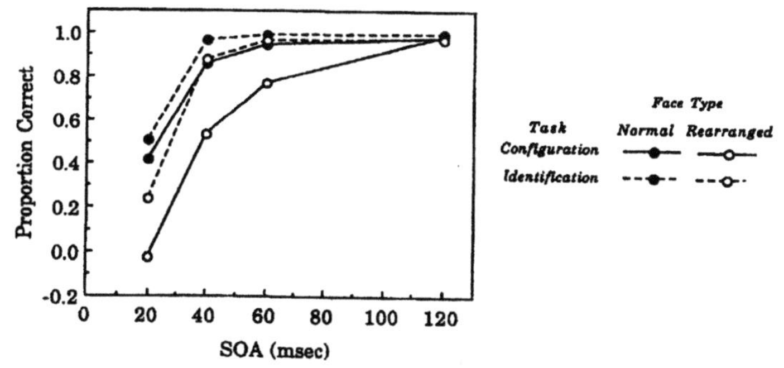

Fig. 4.3. Proportion of correct on the tasks of classification of spatial configuration (i. e., whether the face features were normal or rearranged), and of determination of the identity of the normal or rearranged faces, as a function of SOA. The data has been corrected for guessing. (Redrawn from Rolls ET, Tovée MJ, Purcell DG, Stewart AL, Azzopardi P, Exp Brain Res 1994; 101:473-484.).

Given the comparatively small number of stimuli, the results were corrected for guessing. The results were thus scaled so that chance performance was considered as 0% correct and perfect performance as 100% correct (Fig. 4.3). The performance on the identification task was better than on the configuration task, but this may be because the test faces were all known to the observers.

A further psychophysical task used to investigate the effects of masking was to ask the subjects to rate the clarity of the image on a scale of 1 to 10.[4] This was referred to as the face clarification effect. The procedure was as before, but only two faces and their jumbled versions were used. The mean magnitude estimates given by subjects for the clarity of the faces in normal configuration and with rearranged features are shown in Figure 4.4. For simplicity only the first four SOA values are plotted. The normal face was judged clearer than the rearranged face at all SOA values, but the difference in clarity was reduced as the SOA shortened.

The Effects of Masking on Neural Responses

The neurophysiological experiments were all carried out using the same presentation system and task timing as the psychophysical experiments. The responses of single neurons in IT cortex and walls of the STS were recorded in awake behaving rhesus monkeys. Examples of the effects of backward masking on the responses of a

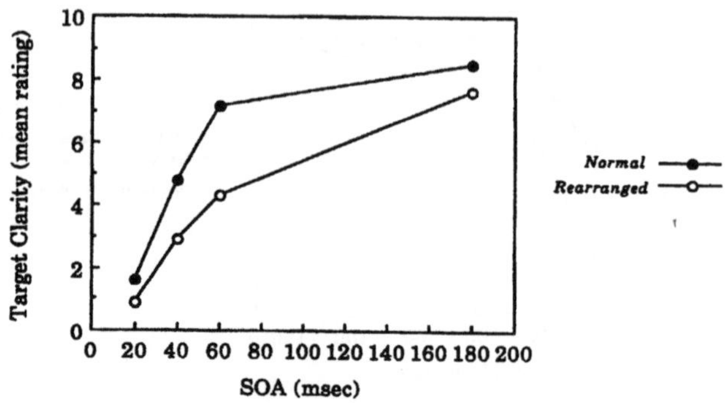

Fig. 4.4. Rating of the clarity of faces when normal or rearranged as a function of SOA. (Redrawn from Rolls ET, Tovée MJ, Purcell DG, Stewart AL, Azzopardi P, Exp Brain Res 1994; 101:473-484.).

single neuron are shown in peristimulus rastergram and time histogram form in Figure 4.5. The rastergram-spike density histogram pair in Figure 4.5a shows the responses of the neuron to a single frame of the test stimulus (an effective face stimulus for that neuron). Relative to the prestimulus rate, there was an increase in the firing produced with a latency of approximately 75 msec, and this firing lasted for 200-300 msec. This is much longer than the 16 msec presentation of the target stimulus. In figure 4.5b the effects of introducing a noneffective face as the masking stimulus with different SOAs are shown. It is shown that the effect of the mask is to limit the duration of the firing produced by the target stimulus. Very similar masking was obtained with the standard N-O pattern mask. Similar experiments were repeated on 42 different cells, and in all cases the temporal aspects of the masking were similar to those shown in Figure 4.5.

The effect of the mask on cumulative firing rate of a single cell is illustrated in Figure 4.6, in which the firing rate under the five SOA conditions is shown. The effect of the mask is to strongly depress the firing, such that for an SOA of 20 msec, the firing rate is only a fraction of what it is at an SOA of 1,000 msec. Moreover, the peak firing rate and the length of time the neuron is active can be titrated down as the SOA is reduced. The data can also be shown as a percentage reduction in firing rate relative to the 1,000 msec SOA, as is shown in Figure 4.7, where the average for 20 cells is displayed.

The data shows a progressive reduction in percentage firing rate as the SOA is reduced. These results have been replicated in a study by Kovacs, Vogels and Orban.[11]

These effects have shown an interruption of the processing of an effective stimulus for the cell, produced by a mask which did not itself activate the cell. The neurophysicial data provide direct evidence that in this case backward masking operates by interruption, but what happens if the mask is a stimulus which also happens to activate the cell? This is illustrated in Figure 4.8, in which an effective stimulus for the cell was presented as a mask at different time intervals after an effective target. The response to the masking stimulus, a face which activated the neuron, was similar to the response of the neuron to the target stimulus, which is shown in Figure 4.8a. The response to the target stimulus was shortened and reduced when it was followed by the effective mask stimulus at short SOAs (as can be seen at 20 msec and 40 msec in Fig. 4.8). Indeed, the neuron's response at these SOAs to an effective test face followed by an effective face stimulus used as a mask was smaller than to the test stimulus presented alone (compare Fig. 4.8a, when only the target stimulus was shown, with the traces for SOAs of 60 msec and 100 msec; there appeared to be two peaks in the neuronal response firing (see Fig. 4.8). This would be consistent with feed forward inhibition (produced by the mask) of the cell's response to the target stimulus, followed by a response to the mask stimulus. Thus, at SOAs of 60-100 msec, in the visual system, at least at the cellular level, there appear to be temporally separate responses to two successive effective stimuli for a cell. Similar effects were found in all five neurons tested in this way, with an effective stimulus for the cell used in a backward masking paradigm.[5]

Information Analysis of the Effects of Visual Masking

The effect of the mask is to progressively reduce both the peak firing rate and the length of time the neuron appears to be active. To quantify the change in information this entails, Bayesian information analysis (see chapter 3) was applied to the spike trains recorded under the different SOA conditions.[6] Figure 4.9 shows the information calculated in discrete samples of 50 msec, under the five SOA conditions averaged across 16 cells. Information is encoded in the cell's response for several hundred msec after the cell starts responding to the stimulus, rising to a peak at 100-150 msec after the onset of

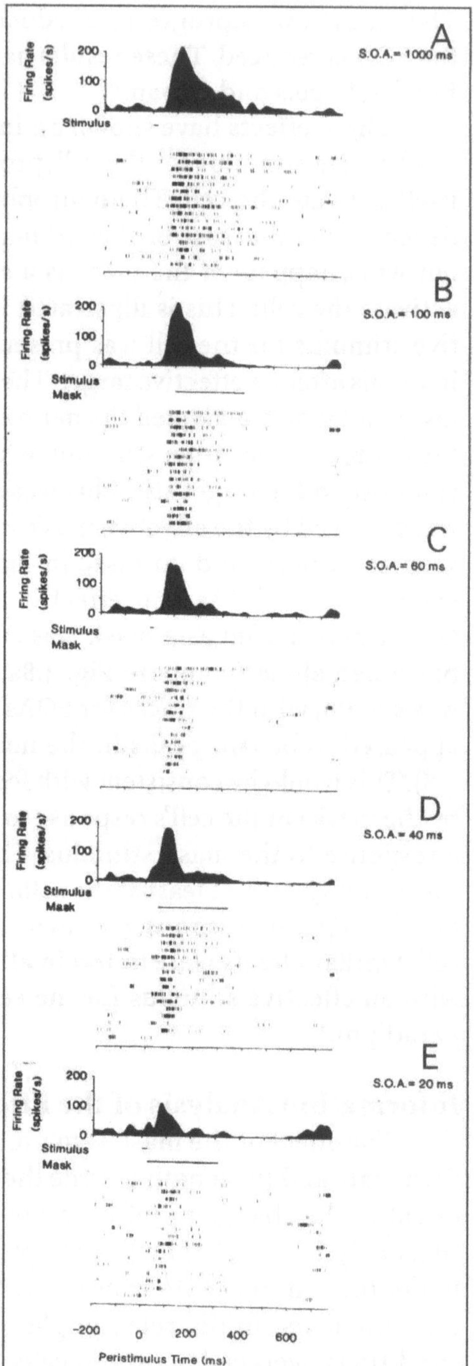

Fig. 4.5. Peristimulus rastergrams and smoothed peristimulus spike density histograms based on responses in 8-16 trials to the test face alone (top raster-histogram pair), and to the test face following by a masking stimulus (which was a face that was ineffective in activating the cell) with different SOA values. The mask alone did not stimulate the cell. The SOA values were (a) 1,000 msec, (b) 100 msec, (c) 60 msec, (d) 40 msec and (e) 20 msec. (Reproduced with permission from Rolls ET, Tovée MJ, Proc R Soc Lond B 1994; 257: 9-15. © (1994) Royal Society).

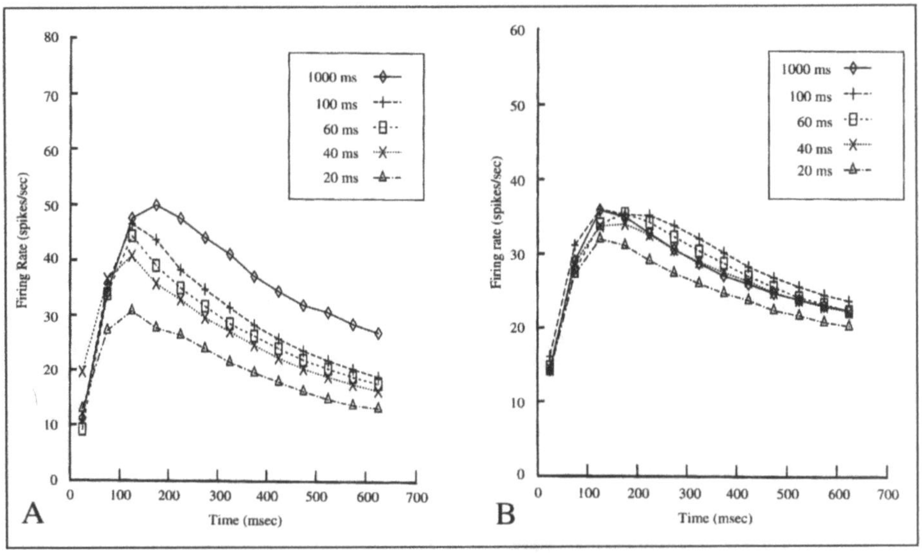

Fig. 4.6. The firing rate calculated in 50-msec samples, under the five SOA conditions for (a) a single neuron and (b) averages from 16 neurons. (Redrawn from Tovée MJ, Rolls ET, Panzeri S, Eur J Neurosci 1998; submitted).

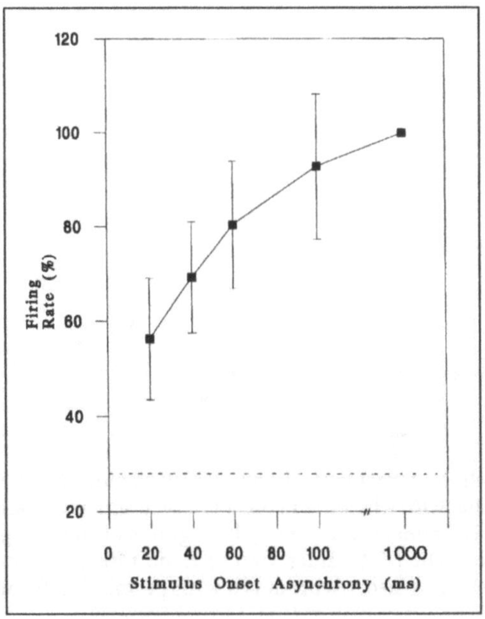

Fig. 4.7. The average response of a population of 20 cells fully analyzed as a function of SOA. The maximal response of each neuron was normalized to 100% without the mask stimulus before the mean percentage response was calculated. The mean and the standard error are shown. The spontaneous firing rate was on average 16% of the maximum firing rate. (Reproduced with permission from Rolls ET, Tovée MJ, Proc R Soc Lond B 1994; 257: 9-15. © (1994) Royal Society).

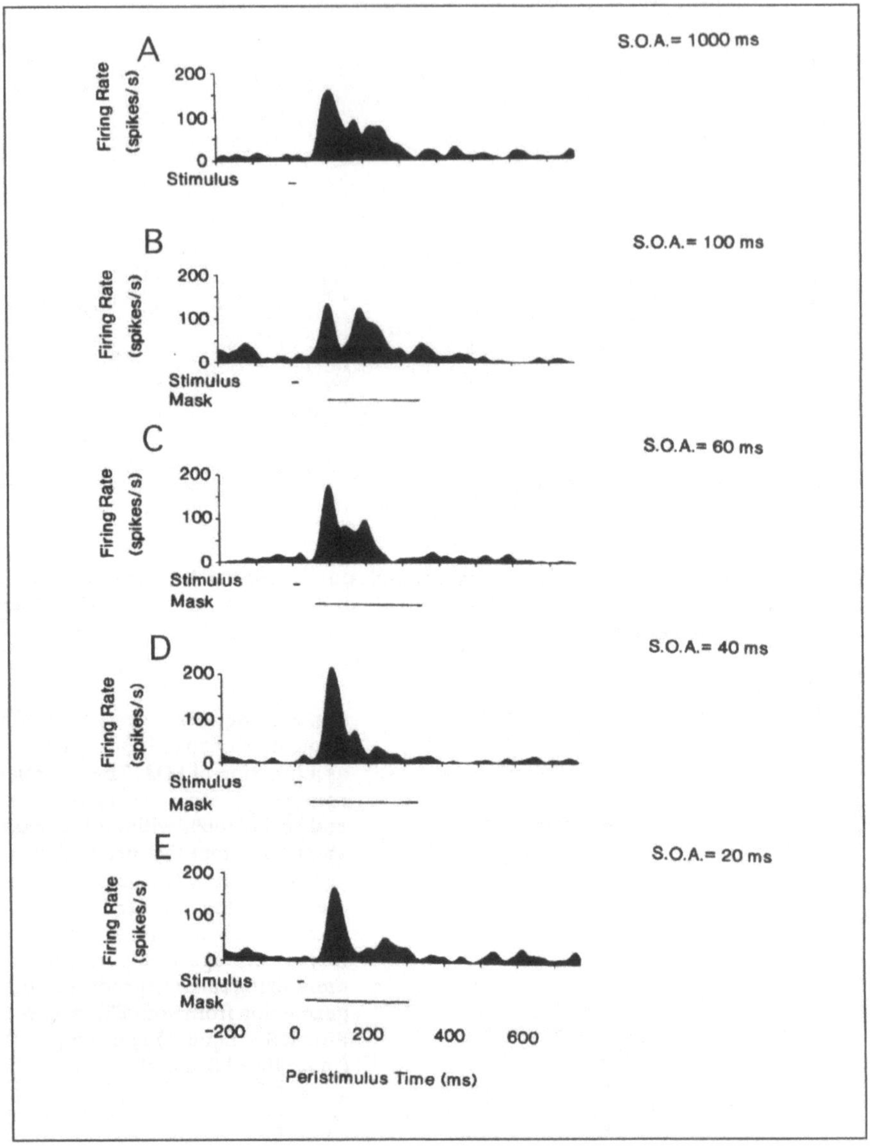

Fig. 4.8. The neuronal responses of a cell to a target face stimulus which was an effective stimulus for the cell as a function of the SOA, when the mask was a face which was an effective stimulus for the neuron. Peristimulus spike density histograms of the firing are shown for each stimulus condition. The SOA values are (A) 1,000 msec, (B) 100 msec, (C) 60 msec, (D) 40 msec and (E) 20 msec. (Reproduced with permission from Rolls ET, Tovée MJ, Proc R Soc Lond B 1994; 257: 9-15. © (1994) Royal Society).

stimulus presentation. The information level then declines rapidly, but the cells continue to encode a significant amount of information about the stimulus for up to 300 msec post stimulus onset. For the other SOA periods, the effect of introducing the mask is to reduce the total information signaled by the cell and to reduce the length of time over which a neuron encodes information. For example, at an SOA of 20 msec, the information peak is reduced to less than a third of the peak reached under the 1,000 msec condition. A similar pattern is shown in Figure 4.10, which illustrates the information from a single cell and averaged across 16 cells calculated from cumulative samples of 50 msec, under the five SOA conditions. Under conditions of a 1,000 msec SOA, the encoded cumulative information rises steadily until 200 msec poststimulus presentation. For the other SOA periods, the effect of introducing the mask is to reduce the total cumulative information signaled by the cell and shortens the period of time over which the information rises. For example, for the 20 msec SOA, the cumulative information is again less than a third of that for the 1,000 msec SOA and it reaches its peak at 150 msec poststimulus presentation, rather than at 200 msec, which is the case for the longer SOAs.

The results of a principal component analysis are consistent with the results detailed above.[6] Principal component 1, which is strongly correlated with firing rate,[12-13] is significantly reduced by the mask. Principal components 2 and 3 were not significantly altered by masking. Previous studies have suggested that these principal components are related to features of the spike train unaffected by the masking paradigm; for example, principal component 2 is believed to be dependent on the relative response latencies of neurons to stimuli.[12-13]

The combination of electrophysiology and information analysis have allowed the quantitative measurement of the change in information encoded by a neuron under conditions of backward visual masking. The effect of the mask is to reduce both the total information signaled by a cell, the size of the information peak and the length of time it is signaling information. These results clearly show that there is a systematic reduction in the amount of information with decreasing SOA. Thus the backward masking paradigm

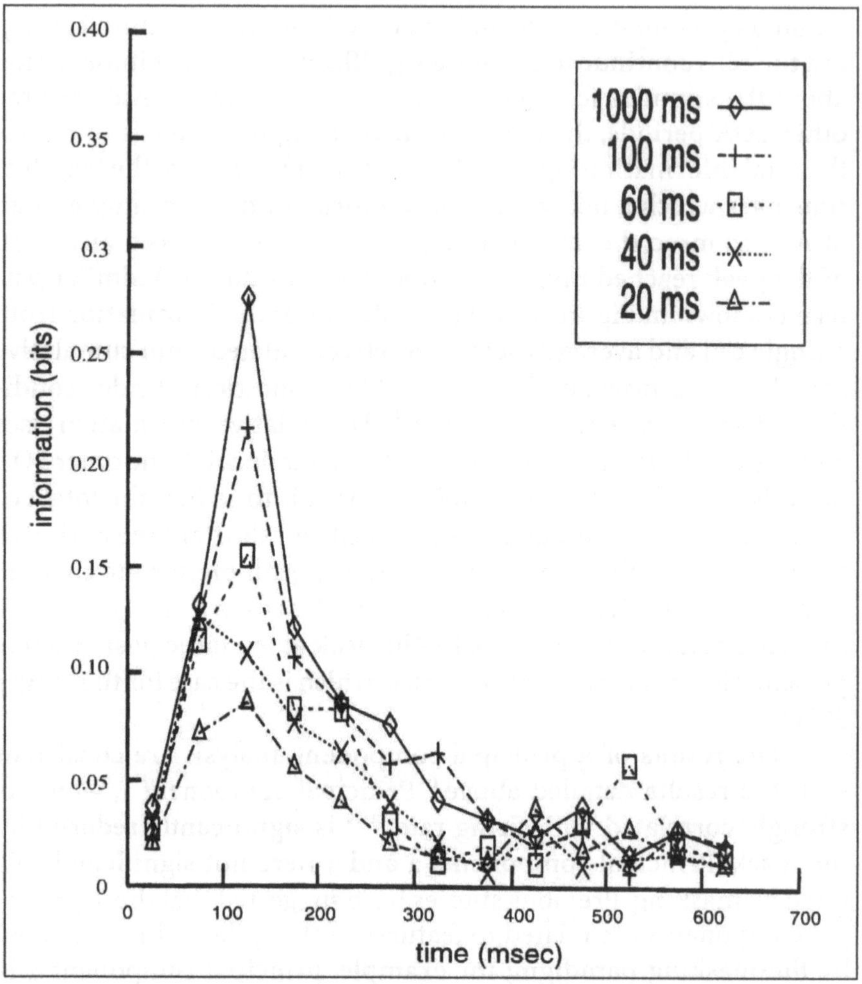

Fig. 4.9. The average information calculated from 16 cells based on discrete 50 msec samples, under the five SOA conditions. (Redrawn from Rolls ET, Tovée MJ, Panzeri S, J Cog Neurosci 1998; in press).

can be used to limit the time available to the visual system to process visual stimuli. By combining this paradigm with a visual recognition task, it is possible to determine the minimum length of time a neuron has to be active to mediate visual recognition.

Neurophysiology of Backward Masking

The neurophysiological and information data described here can be compared directly with the effects of backward masking in human observers, studied in the same apparatus with the same stimuli. For the human observers, identification of which face from

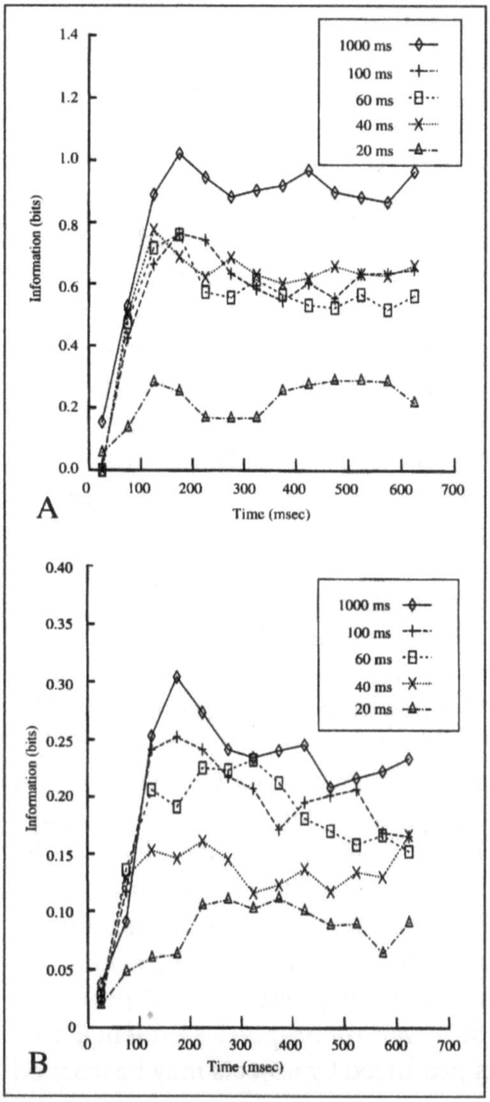

Fig. 4.10. The average cumulative information calculated from cumulative 50 msec samples, under the five SOA conditions, for (a) a single neuron and (b) an average of 16 neurons. (Redrawn from Rolls ET, Tovée MJ, Panzeri S, J Cog Neurosci 1998; in press).

a set of six had been seen was 50% correct with an SOA of 20 msec, and 97% correct with an SOA of 40 msec (see above). Comparing the human performance purely with the changes in firing rate under the same stimulus conditions suggested that when it is just possible to identify which face has been seen, some neurons in a given cortical area may be responding for only 20-30 msec. The implication would be that very short processing times, in the order of tens of milliseconds, are enough for a neuron to perform sufficient computation to enable its output to be used for identification.

These studies also suggest that there is a population of neurons in the temporal cortical areas which respond to a single frame (16 msec) presentation of the target stimulus, a face, with an increased firing rate that can last for 200-300 msec.[4-6] This may represent a form of short-term visual memory based on the continued firing of these neurons after a stimulus has disappeared. This short-term visual memory may be implemented by the connections made between nearby pyramidal cells in the cerebral cortex. These recurrent connections may function in part as an auto-associative memory, which not only enables cortical networks to show continued firing for a few hundred milliseconds after a briefly presented stimulus, but also confer some of the response specifically inherent in the responses of cortical neurons. In this theory, the other mechanism that would make a major contribution to the tuning of a neuron would be the forward synaptic strengths from the preceding stage of processing. One effect which may be facilitated by this short-term visual memory is implementation of a trace learning rule in the visual cortex which may be used to learn invariant representations.[5,14]

The continued neural activity would enable pairwise associations of successive images of the same object, which could form the basis for the construction of an invariant representation of that stimulus. For example, the neurons activated by a stimulus are still in a state of activation (and postsynaptic depolarization) perhaps 500 msec later when the same stimulus is seen translated across the retina, viewed from a different angle, or at a different size, so that the active axons carrying the transformed representation can undergo Hebbian associative synaptic modification onto just those neurons that remain in an activated state from the previous input produced by the same object. In this way, the invariant properties evident across short time epochs of the inputs produced by objects may be learned by the visual system.[14-15]

Face Discrimination and Masking

The effects of masking on the ability of a cell's output to signal the difference between a normal and a rearranged face is shown for a single cell in Figure 4.11. The solid lines show the response to the normal face, and the dashed lines to the face with rearranged features. The top pair of traces show the response of the cell with no masking stimulus. The response of the cell to the normal face was greater than that to the rearranged face. At short SOAs (e.g., 20 msec), the neuron did not respond significantly differently to the normal and rearranged face, but as the SOA was increased through 60 msec the neuron gradually came to respond significantly more to the normal than to the rearranged face. So, based on the firing rate of this cell, it would be possible to differentiate between a jumbled versus a nonjumbled face at SOA values of 60 msec and above.

In Figure 4.12, the mean firing rate and its standard error under each of the stimulus conditions are shown for 3 different cells. The firing rates are measured in a 200 msec period starting 50 msec after the onset of the test face stimulus. These times are used because a typical latency for the start of firing in these cells is a little longer than this, and because humans could not use more than 200 msec of the firing to determine their behavioral response given typical reaction times.[16-17] At an SOA of 1,000 msec, cell AM041 responded to the nonjumbled face with a mean rate of approximately 66 spike/sec (see Fig. 4.12). As the delay between the onset of the mask and the offset of the test stimulus was decreased to 0 msec (an SOA of 20 msec), the neuronal response to the test stimulus decreased to approximately 26 spikes/sec, which was close to the spontaneous rate of 19 spikes/sec. For comparison, there was a smaller response to the jumbled face even with the SOA of 1,000 msec, and this response became less at shorter SOAs until, at 20 and 40 msec SOAs, there was no significant difference between the response to the rearranged and to the normal face.

Cells AM126 and AM129, also shown in Fig. 4.12, showed the same pattern of effects. At short SoAs there was no significant difference between the response to jumbled and nonjumbled faces, whereas there was a significant difference at longer SOAs of 40-60 msec and above. These results mirror the psychophysical findings. At the SOAs at which the cells can begin to discriminate between jumbled and nonjumbled faces, so can the human observers. The response of the neurons can thus be used to predict the performance of whole observers on the masking task. This parallels the findings from single cell recording studies in V5, where the activity of single neurons can be used to predict the behavior of the whole monkey on a motion discrimination task.[18]

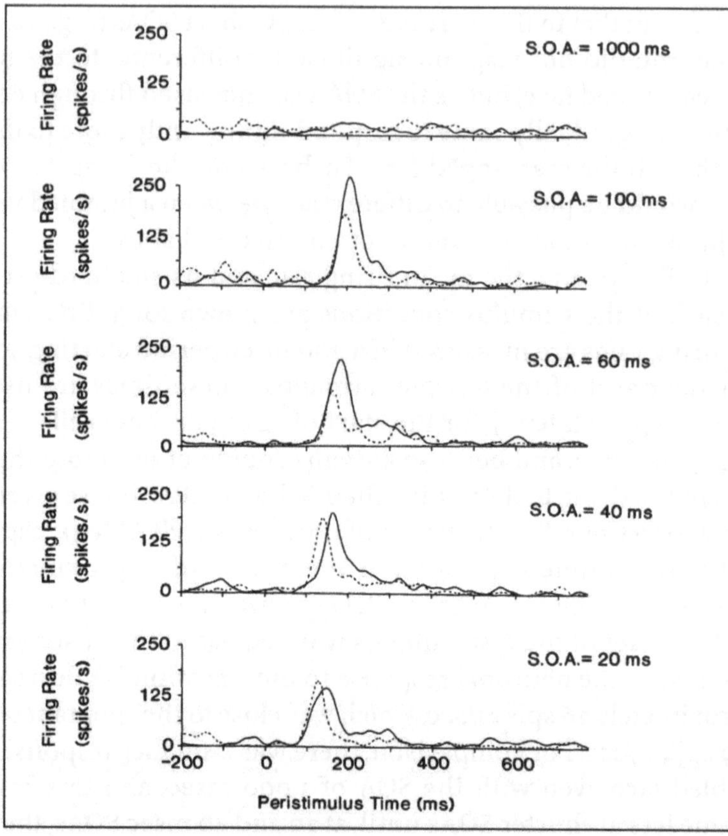

Fig. 4.11. Peristimulus time histograms to show the time course of the masking effects on the responses of a single neuron. The solid lines show the response to the normal face, and the dashed lines to the faces with rearranged features. The test face stimulus appeared at time zero, and its presentation took 16 msec. The top pair of traces show the response of the cell to the test stimulus with no masking stimulus present in the time period shown; as with an SOA of 1,000 msec, the mask did not appear until after the end of the period shown. The peristimulus time histograms are based on 8-10 trials for each SOA (run in random order), and the histograms are Gaussian smoothed with a width σ of 10 msec. (Reproduced with permission from Rolls ET, Tovée MJ, Purcell DG, Stewart AL, Azzopardi P. Exp Brain Res 1994; 101:473-484. © (1994) Springer-Verlag).

Neural Activation and the Face Detection Effect

The degree to which face-selective neurons are stimulated by a face is dependent on a number of factors. Many face-selective neurons are sensitive to whether a face is inverted or not, and whether it is jumbled or not. So an inverted or jumbled face will stimulate less neurons than a normal face. This is supported by evidence from vi-

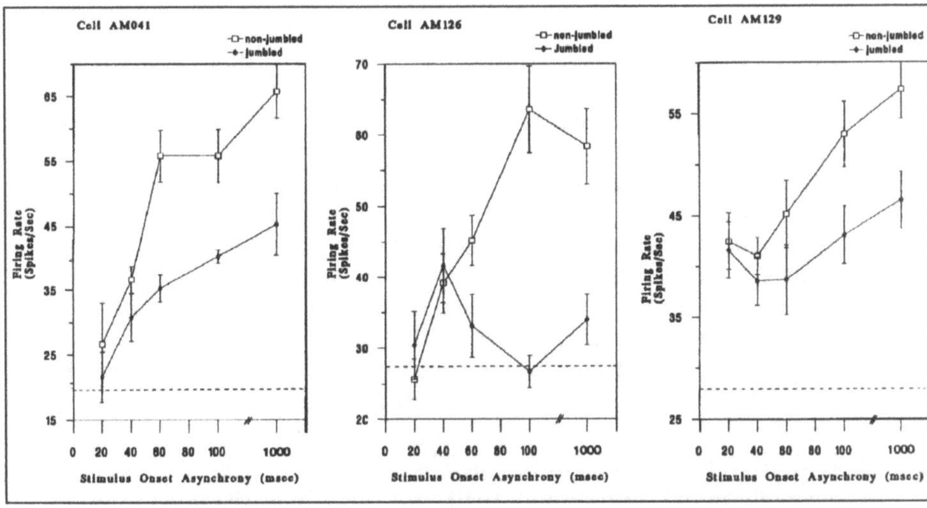

Fig. 4.12. The responses of three different cells (AM041, AM126 and AM129) to the test face stimulus in normal configuration and in the jumbled configuration as a function of SOA. The mean and standard error of the mean firing rate measured in a 200 msec period starting 100 msec after stimulus onset is shown for each SOA value, and the dashed horizontal line shows the spontaneous firing rate. (Reproduced with permission from Rolls ET, Tovée MJ, Purcell DG, Stewart AL, Azzopardi P. Exp Brain Res 1994; 101:473-484. ©(1994) Springer-Verlag).

sual evoked potential studies, where neural activity in the human temporal lobe can be evoked by tachistoscopic presentation of faces.[19-20] Although this response is largely independent of factors such as image size, it is effected by the configuration of facial features. This also suggests that a jumbled face stimulates less neurons than a normal face. So in the case of the FDE, a decision on whether a face has been seen or not will be based on a larger number of activated neurons if the face is in the normal configuration and orientation. This argument can be extended to other psychophysical measures of human performance for faces, including clarity of a face, detection of whether it is rearranged or not, and its identification. All of these can be explained by more face-selective neurons being activated by face stimuli with their features in their normal configuration, so that more 'analyzers' are available with which to make the necessary perceptual judgment. The argument can also be used to explain the recognition of faces, or indeed other objects, from unusual angles. There is a significantly slower recognition for objects and faces if they are viewed at an unfamiliar angle, and this may again be due to the degree of activation of face-selective or object-selective neurons.[21-22]

References

1. Purcell DG, Stewart AL. The face detection effect. Bull Psychonomic Soc 1986; 24:118-120.
2. Purcell DG, Stewart AL. The face detection effect; configuaration enhances detection. Percept Psychophys 1988; 43:356-366.
3. Altmann L, Eckhorn R, Singer W. Temporal integration in the visual system. Influence of temporal dispersion on figure-ground discrimination. Vis Res 1986; 26:1949-1957.
4. Rolls ET, Tovée MJ, Purcell DG, Stewart AL, Azzopardi P. The responses of neurons in the temporal cortex of primates and face identification and detection. Exp Brain Res 1994; 101:473-484.
5. Rolls ET, Tovée MJ. Processing speed in the cerebral cortex and the neurophysiology of backward masking. Proc R Soc Lond B 1994; 257:9-15.
6. Rolls ET, Tovée MJ, Panzeri S. The neurophysiology of backward visual masking: Information Analysis. J Cogn Neurosci 1998; in press.
7. Perrett DI, Rolls ET, Caan W. Visual neurones responsive to faces in the monkey temporal cortex. Exp Brain Res 1982; 47:329-342.
8. Yamane S, Kaji S, Kawano K. What facial features activate face neurons in the inferotemporal cortex of the monkey? Exp Brain Res 1988; 73:209-214.
9. Baylis GC, Rolls ET, Leonard CM. Selectivity between faces in the request of a population of neurons in the cortex in the superior temporal sulcus of the monkey. Brain Res 1985; 342:91-102.
10. Perrett DI, Hietnan JK, Oram MW, Benton PJ. Organization and functions of cells responsive to faces in the temporal cortex. Phil Trans R Soc Lond 1992; 8 335:23-30.
11. Kovacs G,Vogels R, Orban GA. Cortical correlates of pattern backward-masking. Proc Natl Acad Sci USA 1995; 92:5587-5591.
12. Tovée MJ, Rolls ET, Treves A, Bellis RP. Information encoding and the responses of single neurons in the primate temporal visual cortex. J Neurophysiol 1993; 70:640-654.
13. Tovée MJ, Rolls ET. Information encoded in short firing rate epochs by single neurons in the primate temporal visual cortex. Visual Cognition 1995; 2:35-58.
14. Wallis G, Rolls ET. Invariant face and object recognition in the visual system. Prog Neurobiol 1997; 51:167-194.
15. Rolls ET. Neurophysiological mechanisms underlying face processing within and beyond the temporal cortical visual areas. Phil Trans R Soc Lond B 1992; 335:11-21.
16. Thorpe SJ, Imbert M. Biological constraints on connectionist modelling. In: Pfeifer R, Schreter Z, Fogelman-Soulie F eds. Connectionism in Perspective. Elsevier, 1989; 63-92.
17. Thorpe S, Fize D, Mariot C. Speed of processing in the human visual system. Nature 1996; 381:520-522.
18. Shadlen MN, Newsome WT. Motion perception—Seeing and deciding. Proc Natl Acad Sci USA 1996; 93:628-633.

19. Botzel K, Grusser O-J. Electric brain potentials evoked by pictures of faces and non-faces: A search for "face-specific" EEG-potentials. Exp Brain Res 1989; 77:349-360.
20. Jeffreys DA. A face responsive potential recorded from the human scalp potential evoked by faces and by objects. Exp Brain Res 1989; 78:193-202.
21. Perrett DI. View-dependent coding in the ventral stream and its consequences for recognition. In Vision and Movement Mechanisms. In: Caminiti R, Hoffman K-P, Lacquaniti F, Altman J eds. The Cerebral Cortex. Strasbourg: HFSP 1996; 55-63.
22. Mallot H. The view based approach to visual coding. In Vision and Movement Mechanisms. In: Caminiti R Hoffman K-P Lacquaniti F and Altman J eds. The Cerebral Cortex. Strasbourg: HFSP 1996; 133-142.

CHAPTER 5

Population Encoding

What is a Population Code?

The face-selective neurons appear superficially to resemble the gnostic units proposed by Konorski[1] or the cardinal cells proposed by Barlow.[2] These units were described as being at the top of a processing pyramid that began with line and edge detectors in V1 and continued with detectors of increasing complexity until a unit was reached that represented one specific object or person, such as your grandmother, leading to the name by which this theory became derisively known. This idea, which is also called local coding, had two serious problems. Firstly, the number of objects you meet in the course of your lifetime is immense, much larger than the number of neurons available to encode them on a one-to-one basis. Secondly, such a method of encoding is extremely inefficient, as it would mean that there would need to be a vast number of uncommitted cells kept in reserve to code for the new objects one would be likely to meet in the future.

Although individual cells respond differently to different faces, there is no evidence for a face cell that responds exclusively to one individual.[3-4] Face-selective neurons seem to comprise a distributed network for the encoding of faces, just as other cells in IT cortex probably comprise a distributed network for the coding of general object features. Faces are thus encoded by the combined activity of populations or ensembles of cells. The representation of a face would depend on the emergent spatial and temporal distribution of activity within the ensemble.[4-5] Representation of specific faces or objects in a population code overcomes the two most serious disadvantages of the grandmother cell concept. The number of faces encoded by a population of cells can be much larger than the number of cells which make up that population. So it is unnecessary to have a one to one relationship between stimulus and cell. Secondly, no large pool of uncommitted cells is necessary. Single cell experiments have shown that the responses of individual neurons within a population alter to incorporate the representation of novel stimuli within the responses of existing populations.[6-8]

A distributed representation also has the potential to show generalization as the nature of the input changes, and to display graceful degradation or fault tolerance if the network of cells in which the representation is encoded is incomplete or damaged. Another important advantage is the sheer number of stimuli that can be represented by a distributed neural code. Consider the number of stimuli that can be encoded by a population of N neurons if one assumes the absence of noise. If local encoding is used and the neuron shows binary activity (i.e., the neuron is either active or not), then N different representations can be encoded (one neuron is on for each stimulus). If a fully distributed code is used, then 2^N different representations can be encoded (2^N is the number of different combinations of N binary variables).

If the amount of information encoded in a population rises linearly with increasing cell numbers, then the number of stimuli that can be represented rises exponentially. Previous studies have suggested a sub-linear rise in information with increasing cell numbers,[9-10] but this may be related to distortion of the data due to "ceiling effects". That is, the comparatively small number of test images used to stimulate the neurons from which these research groups recorded limits the total amount of information represented by a population made up of these cells. Thus, the information capacity of their "population" is close to saturation almost before they have gone beyond more than a single neuron.

If the neuronal populations can increase the number of stimuli as an exponential of the number of neurons in the population, this would argue strongly for large neuronal populations. If one thinks of the most efficient way a group of neurons could represent stimuli, then one large population will represent far more stimuli than the same number of neurons in two or more populations. These populations would also be able to process information more rapidly, as it takes less time for a unique response to emerge from a large number of units than from a smaller number. There obviously will be limits to how large a population can get, such as the physiological costs of maintaining connections between a large population and the time factor introduced by coordinating such large populations.

Distributed populations have the potential for a very large representational capacity, provided that different cells respond to a different set of stimuli (that is, the set of stimuli that cell A responds to has no stimuli in common with the set of stimuli that cell B responds to: i. e., the activity of the cells is not correlated). This large capacity is achieved if the information encoded by a population of cells increases linearly, in which case the number of stimuli coded grows exponentially with population size. This is in contrast to local representations, in which each stimulus is encoded by a single neuron, so the number of stimuli coded grows only at a linear rate. However, if representations provided by different neurons are strongly correlated, then there would be little increase in representational capacity with population size, no matter how distributed the representations. Large populations would just provide more redundancy.

Pooling the responses of individual neurons into a population response can also be useful in filtering out noise from the signal. However, this assumes that the noise varies independently in each neuron. This may not necessarily be the case. If neurons have similar response patterns and correlated responses, then this suggests a degree of common input, in which case the noise and the signal could vary together and pooling the responses of individual neurons would not reduce the shared noise.[11]

To address these issues, one approach is to examine in detail the response properties of a set of neurons, and determine the effect that their response properties have on how these neurons might function as a population. The study falls into three main sections:

1. Determination of the response properties of a set of face-selective neurons in the primate temporal visual cortex when presented with a large set of test stimuli. This investigation focuses particularly on the response selectivity of the neurons, as this has direct ramifications on the issue of redundancy in cell populations.
2. Based on the response selectivity of the neurons and the amount of information about the test stimuli encoded in the neuronal responses, it is then possible to model how the neurons would function as a population.
3. The number of stimuli that can be shown during an experiment is naturally limited by time constraints. As mentioned above, this limited stimulus set may introduce ceiling or saturation effects in the representation of information by the cell

population. To overcome this difficulty, it is possible to produce simulations of the responses of the neurons to new stimuli, based on the characteristics of the responses to the existing stimulus set. In this way it is possible to model how the population would represent large numbers of stimuli and circumvent ceiling effects.

Neural Response Selectivity

One empirical way of testing how neurons encode visual information in populations is by looking at how selective a neuron is for stimuli both within a class and between stimuli from different classes. A neuron with a highly selective response range would suggest that it is part of a small population, which taken to its extreme would produce local encoding with a one-to-one representation of stimulus to cell, a grandmother cell.[2] A neuron with a lower selectivity would suggest that it is part of a larger population, which taken to its extreme would produce a fully distributed representation in which all the neurons participate.[12-13]

To determine the selectivity of single neurons in the primate temporal cortex, the responses of 14 face-selective neurons from the temporal visual cortex from 2 awake, behaving macaques were recorded to 68 different visual stimuli.[4] The recording sites are shown in Figure 5.1. These stimuli included 23 faces and 45 non-face images of real world scenes (Fig. 5.2). As in previous experiments, the macaques were in primate chairs and performed a visual blink task, with fixation confirmed using the search coil technique.

All the 14 cells showed generally stronger responses to faces than to non-faces (e. g., Fig. 5.3). The importance of the difference in the quality of the responses to faces as opposed to non-faces in discriminating stimuli can be illustrated if one looks at a histogram of the firing rates produced by the face stimuli versus a histogram of the firing rates produced by the non-face stimuli (see Figs. 5.4 and 5.5). The responses to face stimuli cover a wide range of firing rates and the faces are well distributed across this range. However, the responses to the non-face stimuli cover a narrow range of firing rates and the different stimuli within this class are not well differentiated.[4] The precise pattern of responses by each cell to the set of face stimuli differed. Six of the cells showed strongly graded responses, three displayed more weakly graded responses, three showed graded responses which clustered into two or more groups and two of the

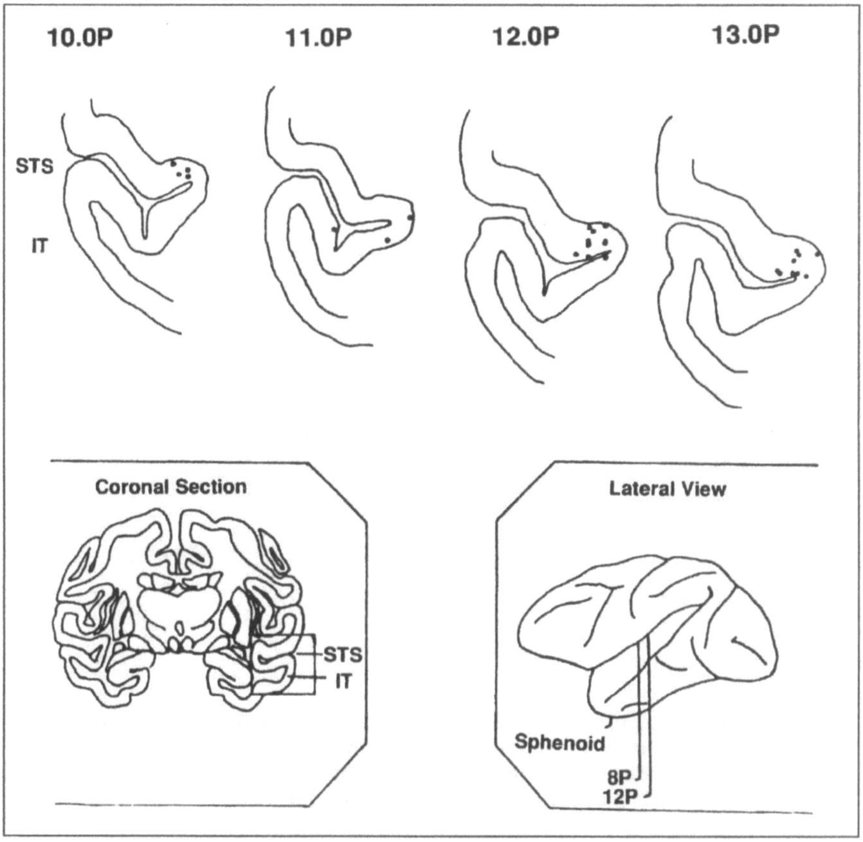

Fig. 5.1. Recording sites for the 14 cells are shown on coronal sections. The positions of the sections are shown on a lateral view of the macaque brain. Distances on the figure refer to mm posterior (P) to the sphenoid reference plane. STS = Superior Temporal Sulcus; IT = Inferior Temporal Cortex. (Reproduced with permission from Rolls ET, Tovée MJ, J Neurophysiol 1995; 73: 713-726. © (1995) The American Physiological Society).

cells showed almost "grandmother cell"-like responses over the limited number of stimuli, with very tight tuning to one or more stimuli (Fig. 5.6).

Application of principal component analysis and Bayesian information measures to the responses of this set of neurons (as discussed in chapter 3) confirmed that their responses contained much more information about which of 20 face stimuli had been seen (on average 0.39 bits) than which of 20 non-face stimuli had been seen (on average 0.07 bits)[4,14] (Figs. 5.7 and 5.8). Even more information

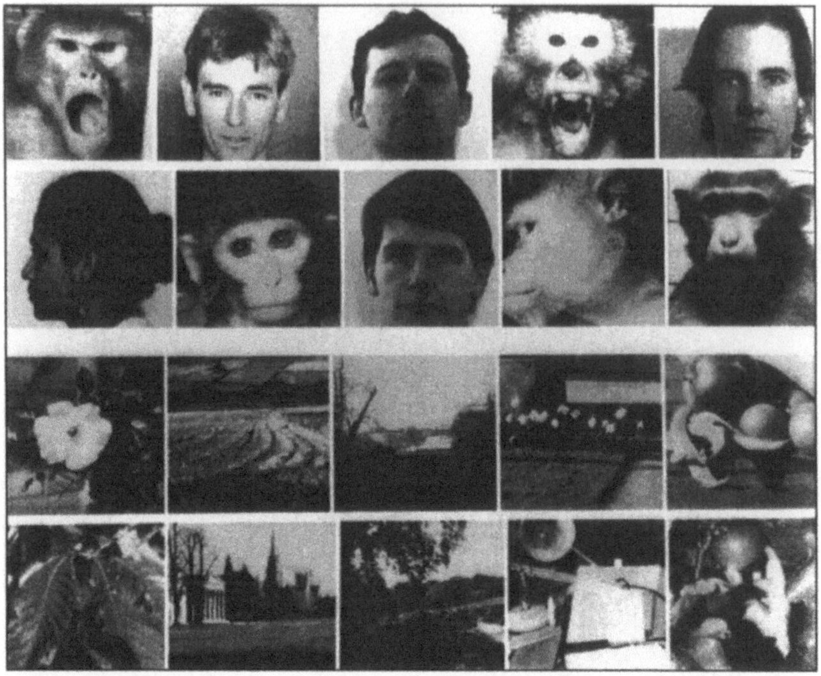

Fig. 5.2. Examples of the 23 face and 45 non-face stimuli used. The non-face stimuli were kindly supplied by Dr. D. Tolhurst of the Physiology Department, Cambridge University. (Reproduced with permission from Rolls ET, Tovée MJ, J Neurophysiol 1995; 73: 713-726. © (1995) The American Physiological Society)

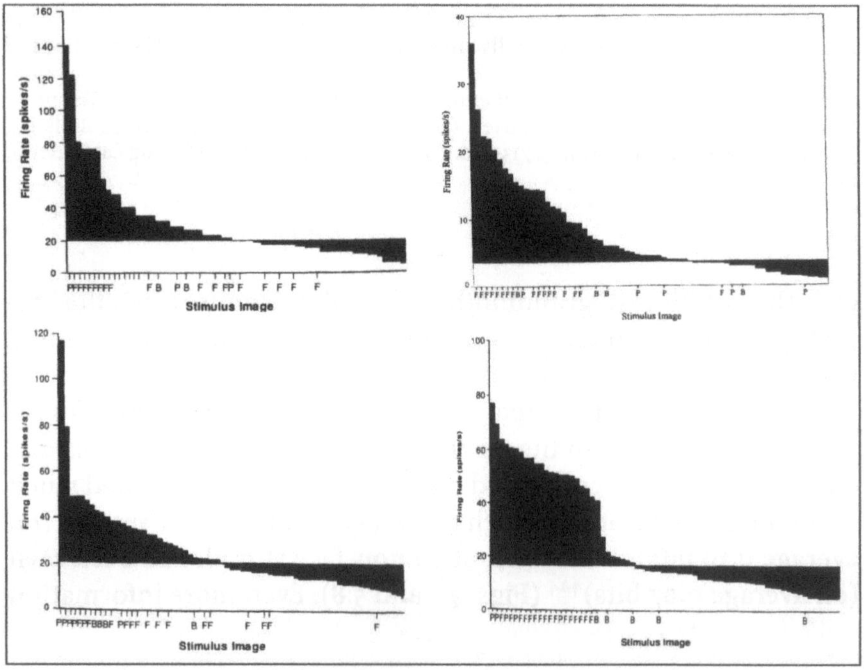

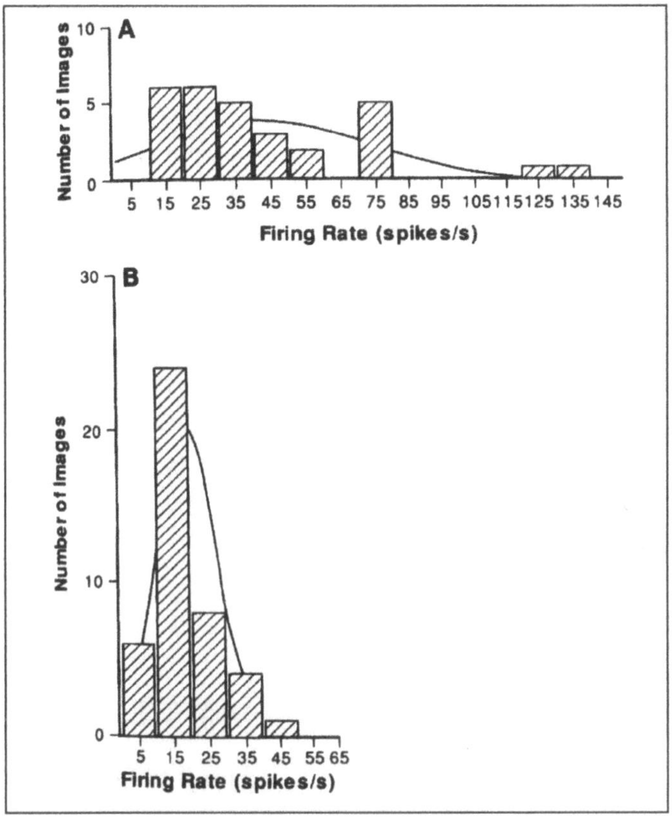

Fig. 5.4. Histograms showing the firing rate distribution produced by (A) the 23 face stimuli and (B) the 45 non-face stimuli for the neuron AM243 (shown in Fig. 5.3). As can be easily seen, the response to face stimuli cover quite a wide range of firing rates, and the different face stimuli are well distributed across this range. However, the responses to the non-face stimuli cover a narrow range of firing rates, and different stimuli within this class are not well differentiated. (Reproduced with permission from Rolls ET, Tovée MJ, J Neurophysiol 1995; 73: 713-726. © (1995) The American Physiological Society).

Fig. 5.3 (opposite). Distribution of firing rates to the 68 stimuli for four single neurons. The firing rate is shown on the ordinate, and the histogram bars are drawn to show changes in firing rate from the spontaneous firing rate (the neuronal "response") produced by each stimulus. Face stimuli that are full faces are labeled F, and those which are in profile are labeled P. Images which included a face or body within a natural scene are labeled B. Stimuli which did not include a face were not labeled. The top left hand histogram shows the responses of a cell (AM243) which was tightly tuned to 2 stimuli out of the set and had a more graded response to additional stimuli. The spontaneous firing rate of the neuron was 20 spikes/sec. The top right hand histogram shows the responses of a cell (AM231) which was more broadly tuned to the stimulus set. The spontaneous firing rate of the neuron was 3.7 spikes/sec. The bottom left hand histogram shows the responses of a cell (AM236) which was very sharply tuned to one face in profile. The spontaneous firing rate of the neuron was 21.0 spikes/sec. The bottom right hand histogram shows the responses of a cell (AM240) which was broadly tuned to the faces in the stimulus set, but responded little to, or was inhibited by, non-face stimuli. The spontaneous firing rate of the neuron was 16.7 spikes/sec. (Redrawn from Rolls ET, Tovée MJ, J Neurophysiol 1995; 73: 713-726.)

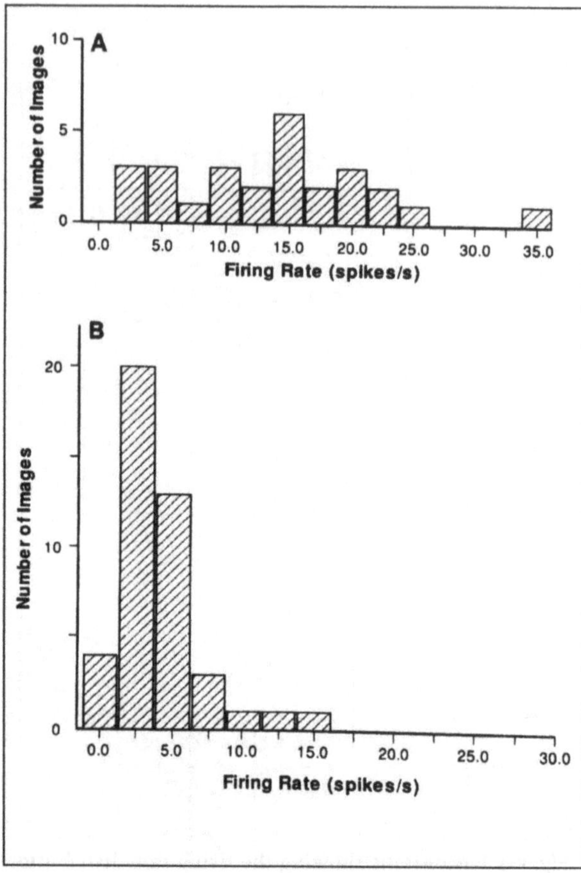

Fig. 5.5. Histograms showing the firing rate distribution produced by (A) the 23 face stimuli and (B) the 45 non-face stimuli for the neuron AM231 (shown in Fig. 5.3). Even though this cell seemed to have a broader range of responses to the images and was less tightly tuned than AM243 or AM236, the responses to face stimuli still cover a wide range of firing rates, and the different face stimuli are well distributed across this range. The responses to the non-face stimuli cover a much narrower range of firing rates, and different stimuli within this class are not well differentiated. (Reproduced with permission from Rolls ET, Tovée MJ, J Neurophysiol 1995; 73: 713-726. © (1995) The American Physiological Society)

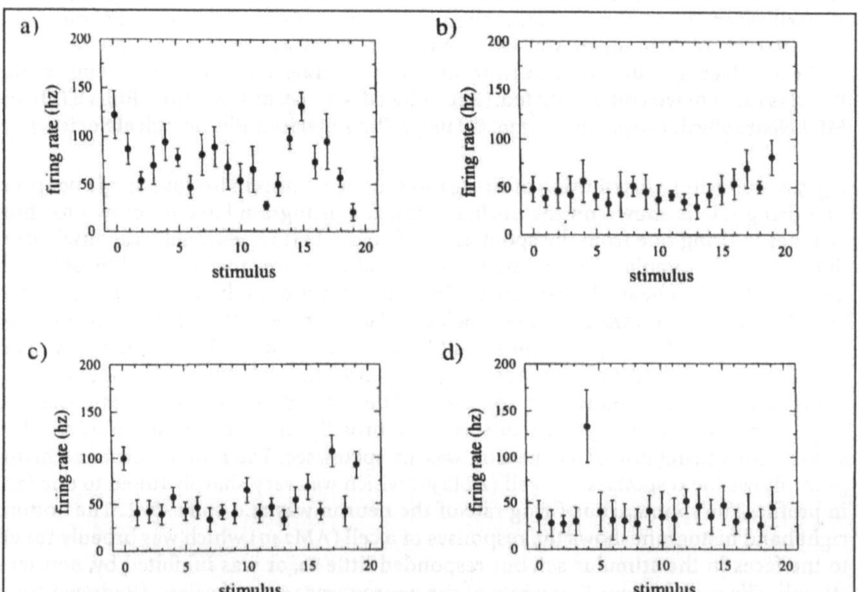

is available per cell when 10 face and 10 non-face stimuli are included in the analysis set (Fig. 5.9). The increased information is mainly due to a wider distribution of firing rates entering the analysis, as the responses to non-face stimuli are very low and those to at least some of the faces are high.

It is also possible to calculate the amount of information I(s,r) available in the neuronal response about each stimulus in the set of s stimuli.[5] As can be seen from Figure 5.10, little information was available from the responses of the neuron to a particular stimulus if the response was close to the average response of the neuron across all stimuli. The amount of information seems to be roughly proportional to how far the neuron's response to a stimulus was from the average response. This was true if the response rate was above, or below, the average response rate. Basically, this means that the response with a low probability, whether it is above or below the mean response rate, will convey the most information. This produces the C-shaped curve shown in Figure 5.10. Although this analysis shows that information is available from responses below the mean response, it does not mean that the cell, or the population network in which it is embedded, can or does utilize this information.

The sparseness of the stimulus representation provided by each of these neurons (which can be thought of as the proportion of stimuli to which that neuron responds) can be defined as

$$a = [\Sigma_{i=1,n} (r_i / n)]^2 / \Sigma_{i=1,n} (r_i^2 / n)$$

where r_i is the firing rate to the *i* th stimulus in the set of n stimuli. The sparseness has a maximal value of 1.0, corresponding to the situation when a neuron responds to all the stimuli in a set of test stimuli. For this set of neurons, the sparseness of the representation of the 68 stimuli by each neuron had an average across all neurons of 0.65 ± 0.16 (Fig. 5.11). The population of neurons had a sparseness measure of 0.82 ± 0.12 to the 23 faces and of 0.60 ± 0.24 to the 45 faces.[4]

Fig. 5.6 (opposite, bottom). The average firing rates and standard deviations to the 20 face stimuli for 4 of the 14 cells. Stimuli are numbered 0 to 19. Rates are in spikes per second. (a) One of six neurons with a highly graded response profile. (b) One of three neurons with a weakly graded response. (c) One of three neurons with graded responses that tended to fall into groups. (d) One of two "grandmother cell"-like cells. (Reproduced with permission from Abbott LF, Rolls ET, Tovée MJ. Cereb Cortex 1996; 6:498-505. © (1996) Oxford University Press).

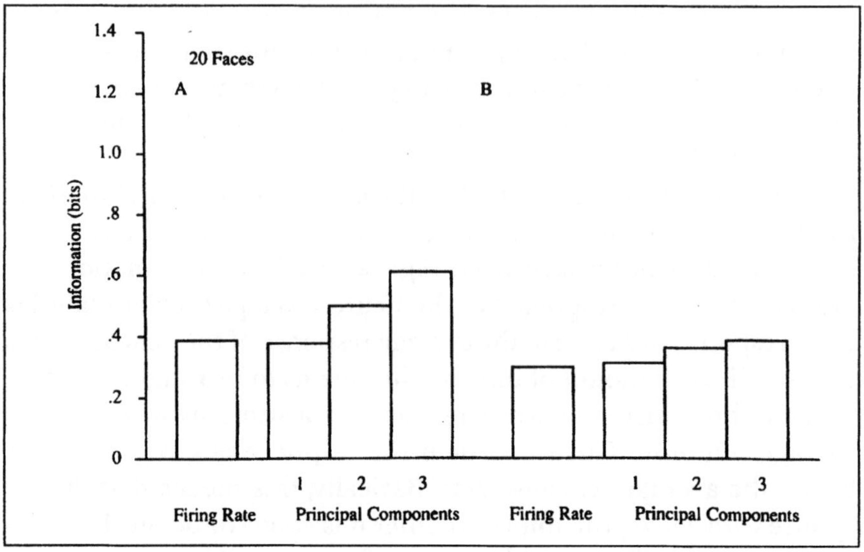

Fig. 5.7. The mean information available in the firing rate and the principal components of each the 14 neurons about the 20 faces in the stimulus set (Principal Component 1 indicates the information available from the first principal component, Principal Component 2 indicates the information available from the first two principal components, and Principal Component 3 indicates the information available from the first three principal components). Figure 5.4A shows the information calculated with correction factor 1 and 5.4B shows the information calculated with correction factor 2. (Reproduced with permission from Rolls ET, Tovée MJ, J Neurophysiol 1995; 73: 713-726. © (1995) The American Physiological Society).

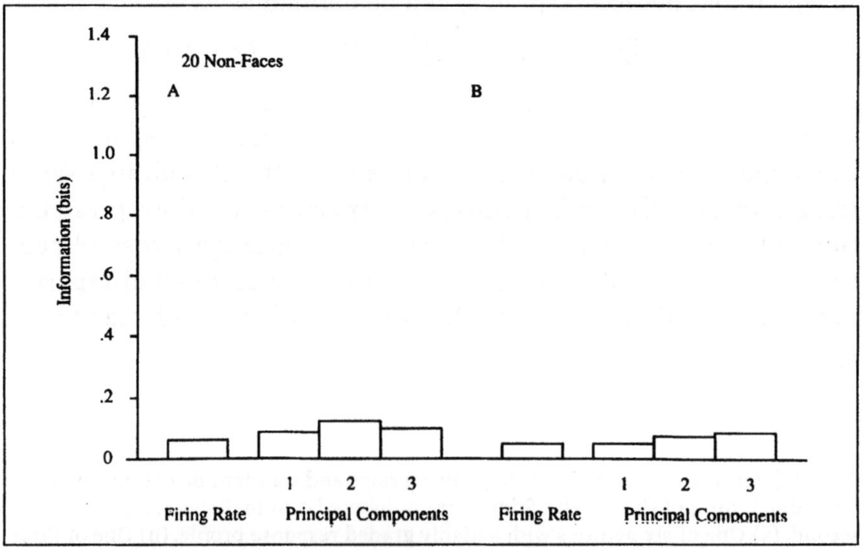

Fig. 5.8. The mean information available in the firing rate and the principal components of each the 14 neurons about the 20 non-faces in the stimulus set. (Reproduced with permission from Rolls ET, Tovée MJ, J Neurophysiol 1995; 73: 713-726. Copyright (1995) The American Physiological Society).

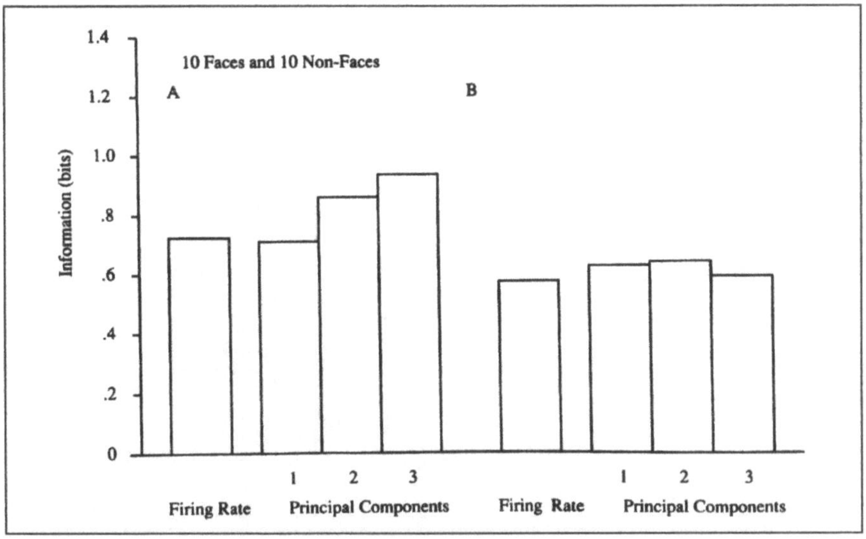

Fig. 5.9. The mean information available in the firing rate and the principal components of each the 14 neurons about 10 of the faces and 10 of the non-faces in the stimulus set. (Reproduced with permission from Rolls ET, Tovée MJ, J Neurophysiol 1995; 73: 713-726. © (1995) The American Physiological Society).

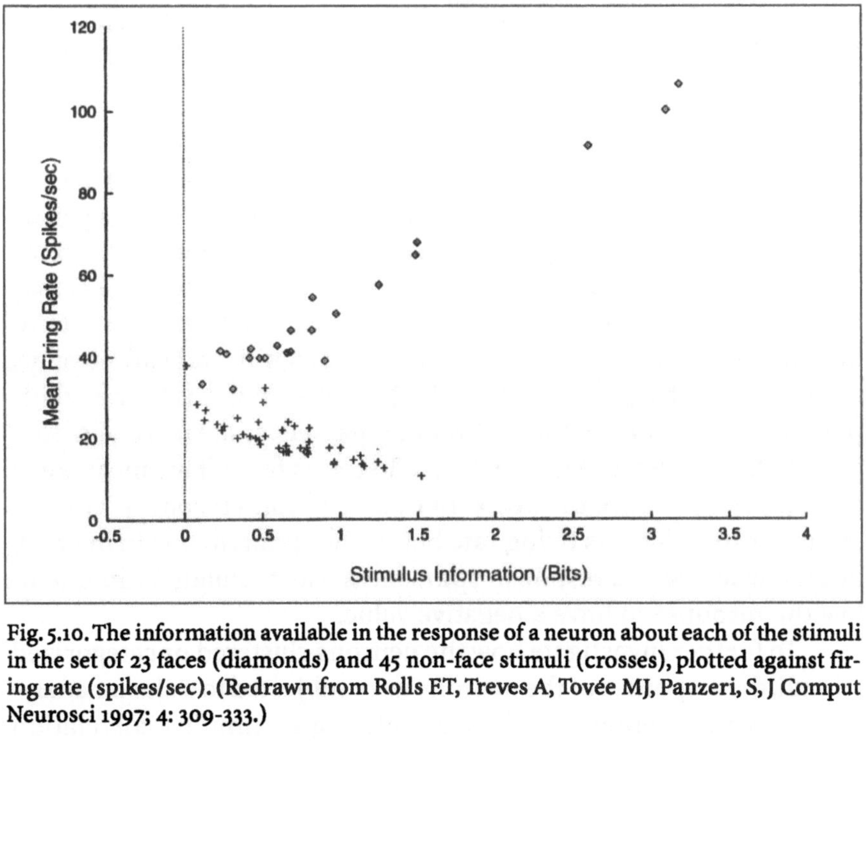

Fig. 5.10. The information available in the response of a neuron about each of the stimuli in the set of 23 faces (diamonds) and 45 non-face stimuli (crosses), plotted against firing rate (spikes/sec). (Redrawn from Rolls ET, Treves A, Tovée MJ, Panzeri, S, J Comput Neurosci 1997; 4: 309-333.)

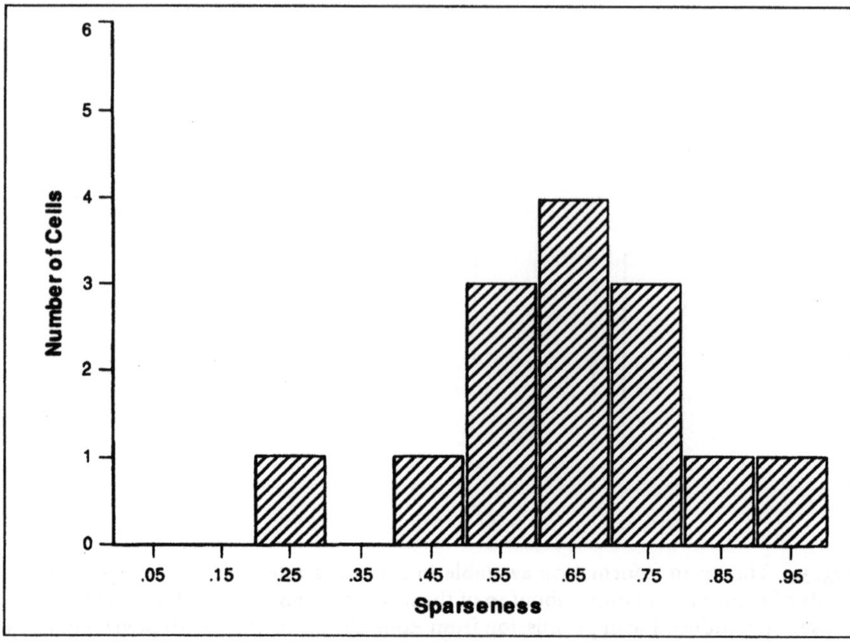

Fig. 5.11. Distribution of sparseness values for the population of cells. The stimulus set over which the sparseness was calculated was the full set of 23 faces and 45 non-face images. The mean sparseness ws 0.63 ± 0.15. (Reproduced with permission from Rolls ET, Tovée MJ, J Neurophysiol 1995; 73: 713-726. © (1995) The American Physiological Society).

The sparseness can also be calculated on the basis of a cell's response rate rather than its firing rate. The response rate is generated by subtracting the spontaneous firing rate of a neuron from the firing rate evoked by a particular stimulus. The response rate is thus the change in the cell's firing rate in response to a particular stimulus. The response-sparseness measure had a significantly lower value, with a mean of 0.33 ± 0.22 for all 68 stimuli (Fig. 5.12). The whole population of neurons had an average sparseness measure of 0.80 ± 0.10 to the 23 faces and of 0.35 ± 0.20 to the 45 faces. This measure of sparseness does have the problem that a neuron's response to some stimuli is to reduce its firing rate below the spontaneous firing rate, but the way that the response-sparseness was calculated did not allow the response to have a negative value.

To further investigate how the neurons clustered or categorized stimuli, the responses of the neurons were subjected to a cluster analysis.[4] This was implemented by calculating a stimulus correlation

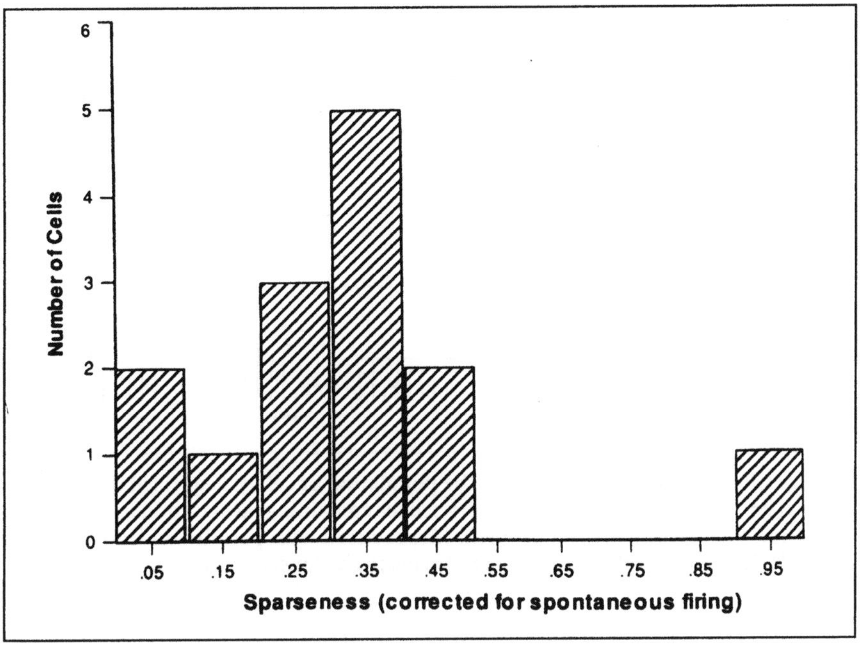

Fig. 5.12. Distribution of response sparseness values for the population of cells. The stimulus set over which the sparseness was calculated was the full set of 23 faces and 45 non-face images. The mean sparseness was 0.33 ± 0.22. (Reproduced with permission from Rolls ET, Tovée MJ, J Neurophysiol 1995; 73: 713-726. © (1995) The American Physiological Society).

matrix, showing the correlation between every pair of stimuli in the responses of the population of neurons. Cluster analysis was then performed on this matrix. The result is shown in Figure 5.13. The first division separated three of the profile views of faces from the other stimuli. Next a juvenile monkey in full-face view was separated from the remaining stimuli. Then many of the remaining face stimuli appeared with different clusterings between the different faces. Within the clusters of 21 stimuli shown on the right of the dendrogram, 19 were of faces, 1 of leaves and 1 of fruit. The next group of five stimuli included all the hands in the set, indicating that some information about hands was provided by this population of neurons. Lastly, all the remaining stimuli to the left of the hands cluster were non-face stimuli and were all clustered close together.

Another way of representing this data is to perform multidimensional scaling (MDS) on the stimulus correlation matrix. MDS produces a configuration of points in a small number of dimensions

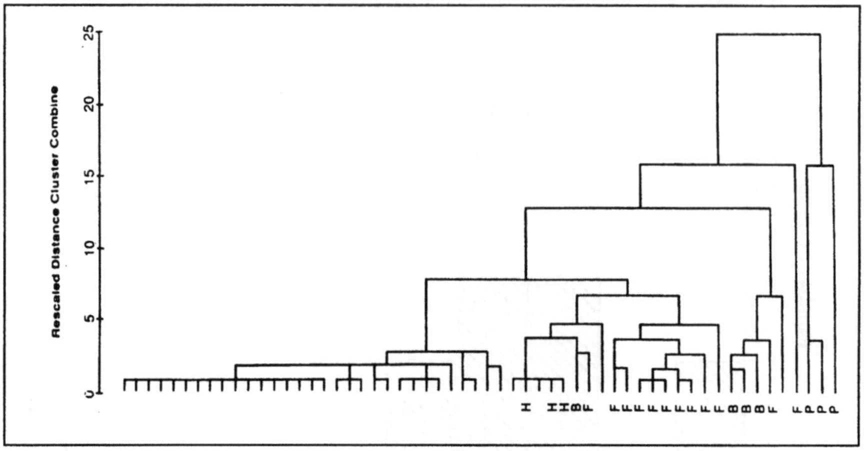

Fig. 5.13. Cluster analysis to show how the population of cells categorized the set of 68 stimuli (P= face in profile, F = full face, B= scenes containing faces or bodies, H= hands. Other non-face stimuli are not labeled). (Reproduced with permission from Rolls ET, Tovée MJ, J Neurophysiol 1995; 73: 713-726. © (1995) The American Physiological Society).

that represent the population responses to the face and non-face stimuli. The distance between the points of the MDS configuration are as close as possible to the Euclidean distances between points corresponding to each face response in a high dimensional space whose dimensions are defined by the cells' responses.[3] Two dimensions accounted for 98% of the variance, and this two dimensional space is shown in Figure 5.14. The different face stimuli were well spread out in this space, consistent with the neurons' responses providing a basis for differentiating different faces. In dimension 1, the profile views of faces were separated from the full face views. The non-face stimuli were not well spread out in the space, but were mostly grouped together at the low end of dimension 1. Some of the 45 "non-face" set contained faces or whole people within the scene, and these stimuli were well spread out along dimension 2. The human and monkey faces were intermixed in the MDS space, and this suggested that the representations provided by these neurons were of differences between faces that would be useful in face identification, rather than a classification of whether the faces were monkey or human.

Young and Yamane[3] also used an MDS analysis on data from populations of face-selective neurons from anterior IT and from STP. The majority of the variance could again be described by 2 dimen-

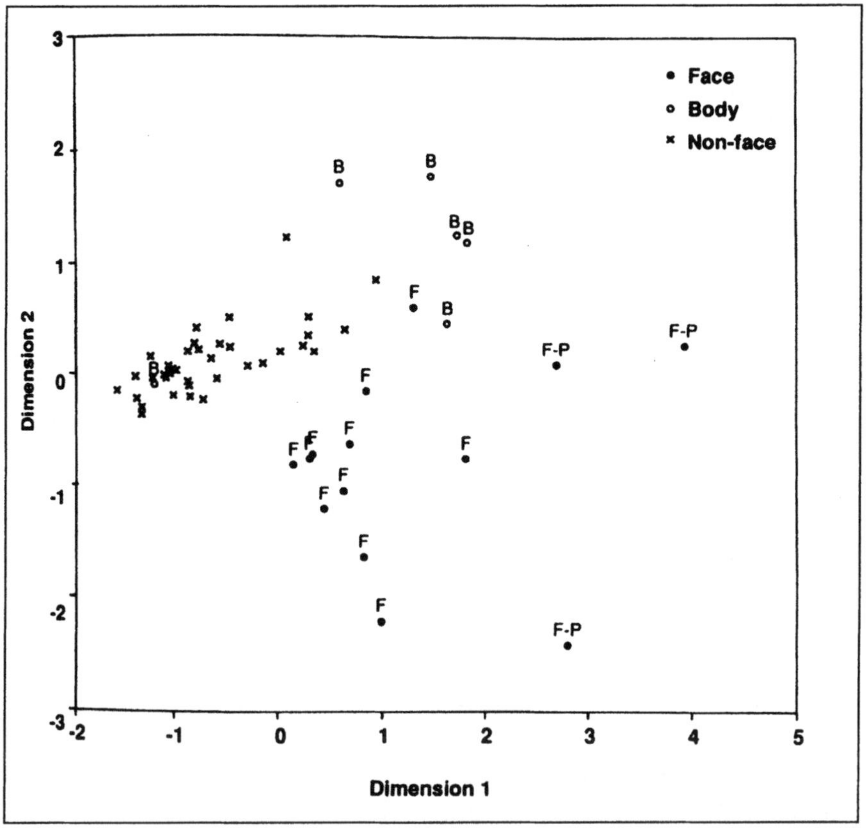

Fig. 5.14. Two dimensional space achieved by multi-dimensional scaling on the stimulus correlation matrix for the 68 stimuli. (F-P = face in profile, F = full face, B= scenes containing faces or bodies. Other non-face stimuli are not labeled). (Reproduced with permission from Rolls ET, Tovée MJ, J Neurophysiol 1995; 73: 713-726. © (1995) The American Physiological Society).

sions. These explained 70% and 75% of the variance from IT and STP cells, respectively. The fact that it is possible to represent the responses of the neural population in only 2 dimensions is a significant dimensional reduction from a much larger number of dimensions (the number of cells in the populations analyzed) to only 2 dimensions, which Young and Yamane suggested indicates a high degree of redundancy in the population code, a point which will be dealt with in detail later in this chapter. However, part of this redundancy may reflect the limited variation in facial physical features in the comparatively small set of faces used in these experiments; 20 faces in the case of Rolls and Tovée[4] and 30 in the case of Young and Yamane.[3] Young and Yamane quantified the physical properties of their face set and applied an MDS analysis to them. A 2 dimensional solution accounted for 94% of the variance, reflecting a great deal of redundancy in the physical information in the face set used. It is therefore perhaps not surprising that the responses of neurons reflect the redundancy of information from the facial features shown. Even so, a PROCRUSTES rotation showed a significant relationship between the physical features of the faces and responses of the population of IT neurons, suggesting that this population may have been coding the general physical features of the presented faces.[3] This was not true of the STP neurons, where there was no significant relationship. This is consistent with previous studies which have suggested that IT neurons are encoding information about face identity, whereas other face-selective neurons in the temporal visual cortex are signaling other information, such as facial emotion or direction of attention.[15-16]

The mean response sparseness suggests that face-selective neurons represent faces with a distributed coding, with any stimulus stimulating a response in up to half the neurons in the population. An obvious criticism of this finding is that the faces in this set are so similar, the face-selective neurons cannot easily distinguish between them. Thus, any given face is likely to stimulate a significant proportion of the cells. However, the distribution of firing rates of individual neurons to individual stimuli, the cluster analysis and the MDS all show that the neurons' responses form a good basis on which to distinguish different face stimuli. These results suggest that the responses of individual neurons to the set of face stimuli are systematically graded, even to faces that are sub-optimal stimuli for a given cell. This result is supported the application to face-selective neurons of a method of vector analysis developed by Georgopoulos[17-18] for neurons in the motor cortex. Young and Yamane[3] applied this technique both to their population data and to the data based on the physical features of their face stimuli. The distribution of the angular discrepancies between the population vectors and the stimulus vectors suggest that the responses of these neurons were indeed graded across their set of face stimuli, responding in a reliable and systematic way to sub-optimal stimuli.

Information from Population Vectors

To further explore how the 14 neurons might function as a population, it is possible to calculate the mean response of each neuron to each of 20 face stimuli from 10 presentations of each stimulus.[4,19] It is then possible to calculate from these population response vectors the relative likelihood that it had been elicited by each stimulus in the set.[5,19] By comparison with the actual stimulus, it is possible to calculate the mean percentage correct identification, as well as the information (in bits) carried about the stimulus by the whole population of neurons on a single trial. When the decoding algorithm used in this calculation approximated to an optimal Bayesian estimate of the relative likelihood, the percentage correct increased from 14% correct (chance was 5% correct) with one neuron to 67% with 14 neurons. The information increased approximately linearly from 0.33 bits with one neuron to 2.77 bits with 14 neurons. Thus the number of stimuli that can be encoded in a population of neurons increases approximately exponentially as the number of cells in the sample increases (in that the log of the number of stimuli increases almost linearly). This is in contrast to a local encoding scheme, where the number of stimuli encoded increases linearly with the number of cells in the sample.

There was some redundancy in the representation of the neurons, with each neuron in this population contributing only 60% of the information it carried rather than 100%; however, this may be at least partly due to the limited stimulus set (only 20 images). This data is consistent with minimal redundancy for sufficiently large and diverse sets of stimuli.

The deviation from linearity seems to be due to a simple ceiling effect. The limited number of stimuli (20 faces) means that the maximum information that can be encoded is only $\log_2 (20) = 4.32$ bits. Near the information ceiling, adding additional neurons to the existing population does not increase the amount of information encoded in the cell population in a linear fashion due to an increased overlap in the response function (i. e., the range of stimuli to which the cell responds) between the new cell and the cells already in the population. The effect of the information ceiling can be shown by altering the level of the ceiling. If the position of the ceiling is varied, the occurrence of the non-linearity is also varied. So for example, if the information ceiling is lowered by reducing the number of stimuli, the nonlinearity occurs earlier. If each cell carries a smaller proportion of the total information, then values close to saturation are approached only for larger populations, and the increase in information is much more linear as cells are added to the sample. This is shown in that by reducing the analysis period to 50 msec, the information available from one cell was 0.13 bits (and 9.4% correct for the 20 stimuli) and the information increased linearly to 1.49 bits (41%) as the number of cells in the population increased to 14 neurons (Fig. 5.15).

Information Encoded by the Cell Population

An additional way of quantifying the potential amount of information a cell population can encode is by using firing rate probability distributions. The analysis is based on a determination of the probability that a particular stimulus, s, evokes a set of firing rates, *r*, in the recorded neurons. This probability can therefore be denoted as P(*r*/s). The breadth of the probability function for a given neuron and stimulus indicates how noisy the encoding is. The stimulus selectivity of a neuron can be derived from the relative probability distributions to different stimuli, and comparing the degree of overlap in this selectivity between different neurons yields the degree of redundancy. These factors are interrelated, and an analysis which takes all these factors into account is Shannon's mutual information. The mutual information is a function of the conditional response probability P(*r*/s) given by:

$$I_r = \sum_{s,r} P(s)P(r/s)\log_2(P(r/s)/P(r))$$

where the sum is over all stimuli and all possible responses. In this equation, P(s) is the probability of a particular stimulus appearing, which is one over the number of stimuli. P(*r*) is the probability of the response set r, which is the sum of P(s)P(*r*/s) over all stimuli.

The information measures were corrected by performing Monte Carlo studies of the results on simulated data where the exact answers were known. The raw and corrected information measures provide upper and lower limits for information content. The information from single neurons varied from 1.0 to 0.1 bits. The average raw information across the population was 0.67 bits, and for the corrected data it was 0.47 bits. The information rises with the number of cells in a way which again seems to be sublinear (see Fig. 5.16). The number of stimuli that can be represented to a given degree of accuracy is proportional to the base-two exponential of the information. The number of faces that can be represented by N neurons grows as a power of N and is somewhere between $1.1N^{0.75}$ and $1.4N^{0.9}$.

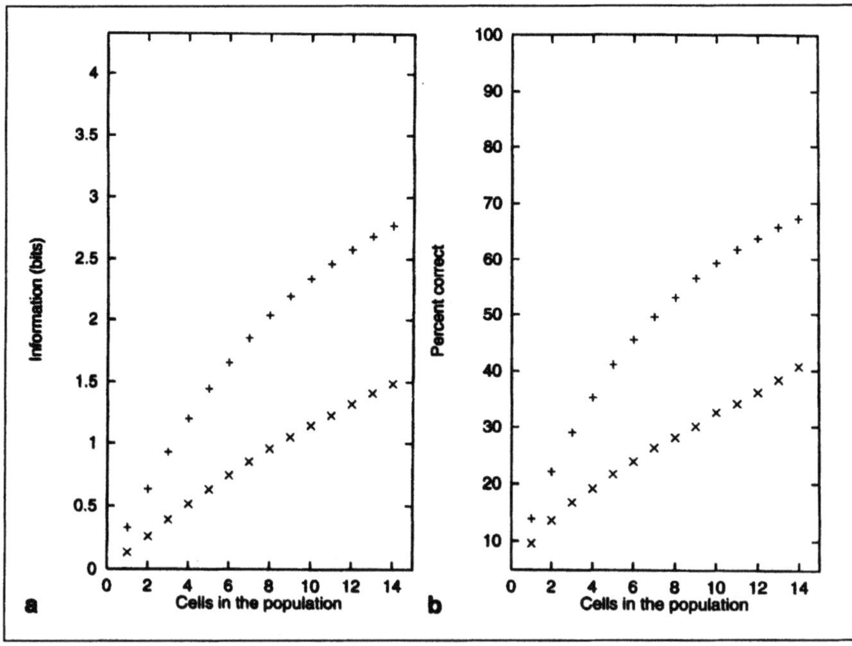

Fig. 5.15. (a)The values for the mean information available in the responses measured in a 500-msec period (+), and in a shorter 50-msec period (x), both starting 100 msec after stimulus onset. The decoding method was probability estimation. (b) The percentage correct for the corresponding data to those shown in (a). (Reproduced from Rolls ET, Treves A, Tovée MJ. Exp Brain Res 1997; 114: 149-162. © (1997) Springer-Verlag).

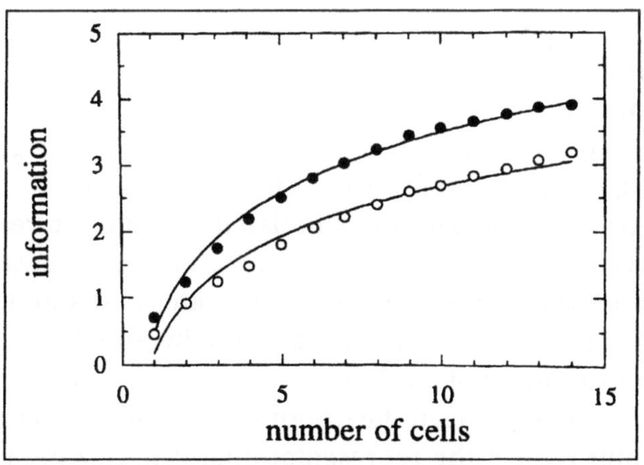

Fig. 5.16. The information for random subsets of different numbers of cells. The curves are logarithmic fits to the plotted points given by $I_{raw} = 0.47 + 0.9 \log_2(N)$ and $I_{cv} = 0.18 + 0.75 \log_2(N)$. (Reproduced with permission from Abbott LF, Rolls ET, Tovée MJ, Cereb Cortex 1996; 6:498-505.© (1996) Oxford University Press).

How Good Is the Code?

Decoding methods were also used to determine how accurately the firing responses could distinguish between different stimuli and how many stimuli could be represented by such responses.[20] Decoding is a procedure for determining which stimulus evoked a particular set of firing rates.[21-22] It can be used to determine the accuracy with which stimuli are represented by neuronal firing. The discrimination accuracy is the percentage of times that a decoding algorithm is able to extract the correct stimulus from a set of rates. To calculate the discrimination accuracy, the probability distributions were used to generate firing rates corresponding to a given stimulus. A decoding procedure was then used to determine which stimulus was most likely to have produced this particular set of rates. Finally, the decoded stimulus was compared with the actual stimulus used to generate rates and, repeating the procedure, computed the percentage of correct decodings. The coding accuracy increases steadily with the number of cells, and it eventually comes close to the ceiling of 100% (see Fig. 5.17). As in the previous measures of information, the rate of increase in the discrimination accuracy may reflect the ceiling rather than the dependence of coding on the number of cells.

Simulating Neural Responses to New Stimuli

To manufacture descriptions of the responses of neurons to new simulated stimuli, a mean firing rate and a firing rate variance for each of the 280 possible cell and stimulus combinations by averaging over all trials was computed. From these means and variances it was possible to construct Gaussian distributions for each cell describing the probability that a particular face evokes a given firing rate. The quality of the fits were checked with the Kolmogorov-Smirnov test. To calculate the simulated stimuli, the responses to additional simulated stimuli were assumed to have Gaussian statistics. So to produce them, average firing rates and standard deviations were generated corresponding to new hypothetical stimuli (see Fig. 5.18 for examples of 4 cells).

Correlations between the responses of different neurons to the same stimuli reduce the representational capacity because of the resulting redundancy. To measure the redundancy of the responses, the Pearson product-moment correlation matrix of the average responses of the different neurons to the same stimulus, over all stimuli, was computed.[20] Off-diagonal elements, indicating correlations, were

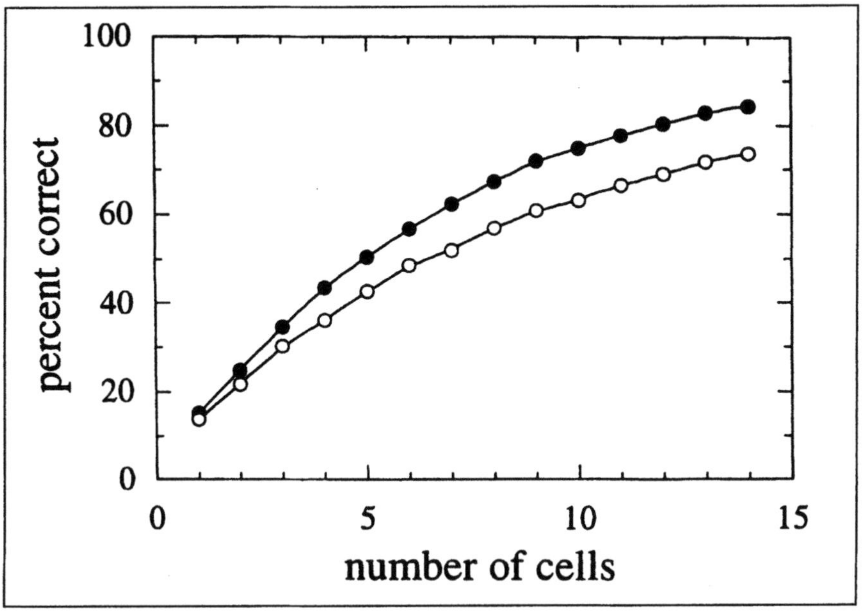

Fig. 5.17. The discrimination accuracy for random subsets of different numbers of cells responding to 20 face stimuli. The upper curve is the raw result and the lower curve the cross-validated result. Discrimination accuracy is defined as the percentage of correctly decoded responses using the optimal maximum likelihood method. (Reproduced with permission from Abbott LF, Rolls ET, Tovée MJ, Cereb Cortex 1996; 6:498-505. © (1996) Oxford University Press).

as large as 0.9, although most tended to lie between +0.5 and -0.5. This again suggests that the representation of faces provided by this set of neurons shows significant redundancy.

The information and discrimination capacity calculation can then be re-run for the set of 14 neurons using large numbers of stimuli (Fig. 5.19). From these results it is clear that the limited size of the stimulus set was the limiting factor in the rise of information as the number of neurons increased.[20] For large numbers of stimuli, the curves illustrated in Figure 5.19 approach a straight line, for which the raw information corresponds to 0.5 bits per cell, and for the cross-validated information is 0.35 bits per cell. This is consistent with the decoding analysis which showed that, at a 50% discrimination accuracy (half the decoded stimuli matched the true stimuli), the number of stimuli that can be represented to a given level of accuracy increases exponentially with the number of coding neurons. This is also true for other discrimination levels. The exponential fits in Figure 5.20 agree with results from the information analysis. The solid

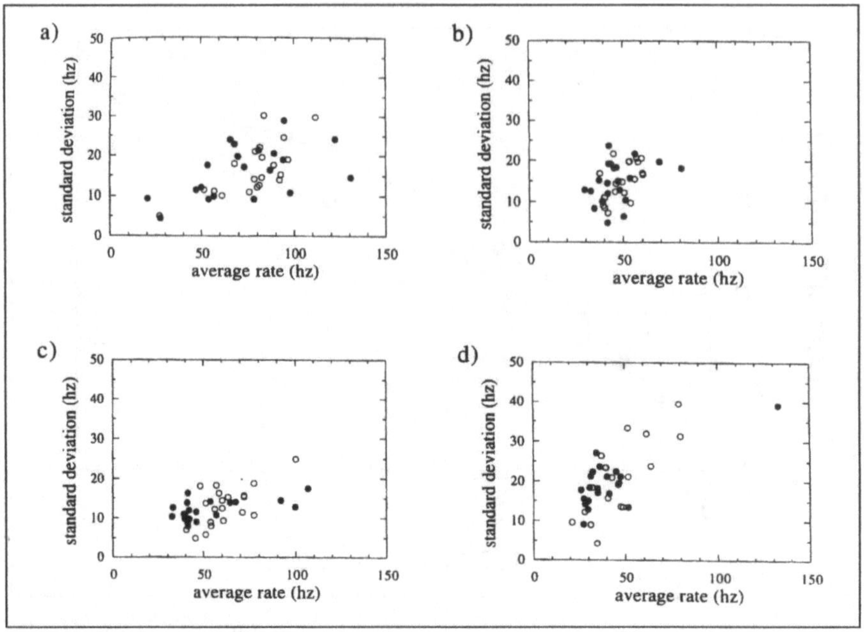

Fig. 5.18. Scatter plots of average firing rates and standard deviations for the twenty original stimuli and twenty simulated stimuli. Solid circles are the actual stimuli in the data set and open circles are the simulated points. (Reproduced with permission from Abbott LF, Rolls ET, Tovée MJ, Cereb Cortex 1996; 6:498-505. © (1996) Oxford University Press).

curve fitting the data points for the raw correct percentage is $2.9(2^{0.47N})$, while the curve for the cross-validated case is $2.9(2^{0.36N})$. These fits indicate that 0.47 and 0.36 bits of information are carried per cell in these two cases, results which are in agreement with the data shown in Figure 5.20, indicating 0.5 and 0.37 bits per cell.[20]

Implications of Distributed Coding

If one chooses a number between the raw and the cross-validated results, it is possible to estimate that N neurons can represent about $3(2^{0.4N})$ faces with 50% discrimination accuracy.[20] This means that the 14 neurons could potentially represent up to 145 faces with this accuracy. The population as a whole carries about 0.4 bits of information about the stimuli per cell. The average information carried by each neuron considered individually was 0.6 bits, showing that redundancy has reduced the information by a third per neuron when they are considered as a population. This redundancy has a

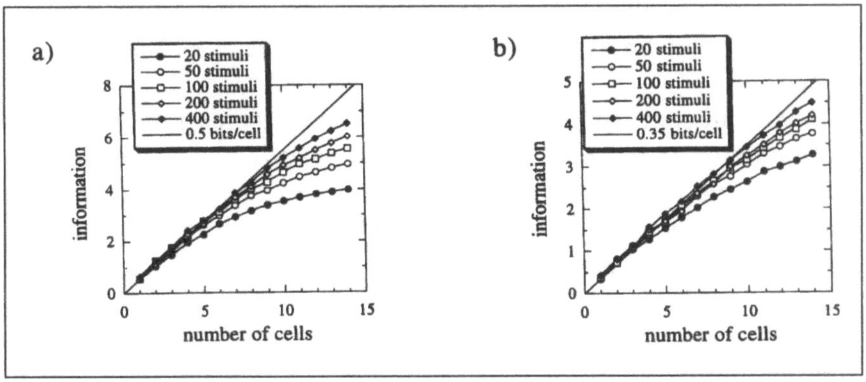

Fig. 5.19. The information as a function of the number of cells computed for different numbers of stimuli. (a) The raw information is plotted for 20, 50, 100, 200 and 400 stimuli. The straight line corresponds to 0.5 bits per neuron. (b) The cross-validated information is plotted for the 20, 50, 100, 200 and 400 simulated stimuli. The straight line corresponds to 0.35 bits per neuron. (Reproduced with permission from Abbott LF, Rolls ET, Tovée MJ, Cereb Cortex 1996; 6:498-505. © (1996) Oxford University Press).

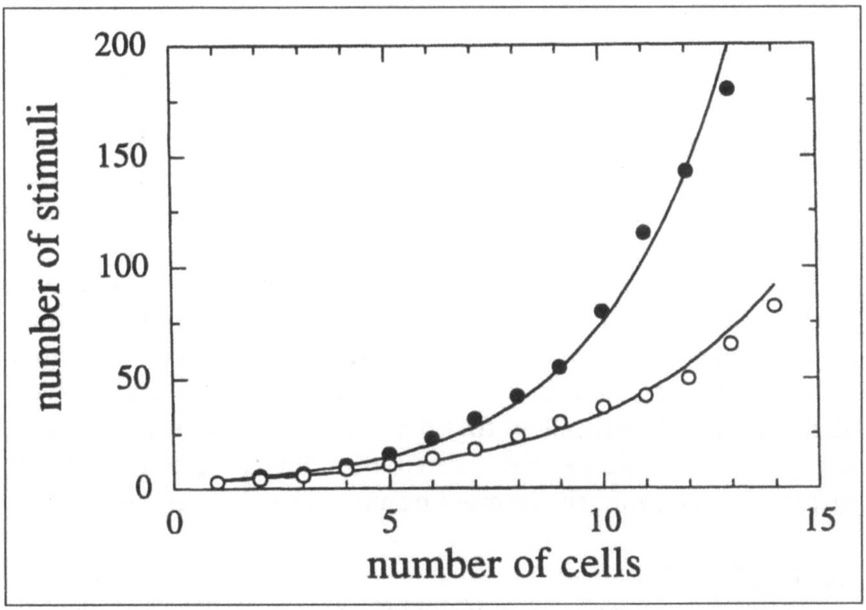

Fig. 5.20. The number of stimuli that can be decoded at 50% accuracy as a function of the number of coding neurons. The solid circles are the result of the raw calculation and the open circles correspond to the cross-validated case. The curves are exponential fits to the data points. (Reproduced with permission from Abbott LF, Rolls ET, Tovée MJ, Cereb Cortex 1996; 6:498-505. © (1996) Oxford University Press).

large impact on the representational capacity of the neuronal population, but may be useful for combating the effects of uncorrelated noise.

Abbott et al[20] suggested a simple model to illustrate the results of this redundancy. If one considers that neurons respond in only two possible ways, such that a fraction (f) of the stimuli elicit one response, while the remaining fraction (1-f) elicit the other response, the information for a neuron will be $-f \log_2(f) - (1-f)\log_2(1-f)$. The value of 0.6 bits for the average information in single neuron responses corresponds to $f = 0.15$ and the 0.4 bits per neuron across the population gives $f = 0.08$. This suggests that the responses of individual neurons distinguish about 15% of the stimuli from the others, but in the full population about half of the 15% distinguished by one neuron are distinguished by the other neurons as well.

A very distributed coding allows the maximum information about a set of stimuli to be provided by a population of neurons, assuming that the degree of correlation in activity is low. Such a distributed code is good for discriminating between stimuli, because the maximum information suitable for comparing fine differences between stimuli would be made available across the whole population. However, a representation as distributed as this might not be optimal for memory storage, where storage capacity may be more important than discrimination.[4-5] The maximum number of firing patterns that can be individually retrieved is proportional to the number of associatively modifiable synapses by a factor that increases roughly with the inverse of the sparseness of the neuronal representation.[23-25] Thus, in the hippocampus, sparseness values are in the order of 0.01-0.014.[25-27] The sparseness of representation thus changes with function. In a sensory processing system, a relatively distributed code will allow fine discrimination of related stimuli within a class, while a sparser distribution may be used in memory systems to maximize the number of memories that can be stored.

Ceiling Effects

Modifying the position of the information ceiling by either altering the number of stimuli the cells encode (either reducing the number included in the analysis or increasing them through simulation) or by reducing the temporal window over which information is calculated has a marked effect on deviation from linearity. That this deviation is due to a ceiling effect is also supported by the good fit to the data provided by the simple model proposed by Gwane and Richmond,[9] which requires as a fitting parameter only mean information from single cells (the initial slope in the rise). In the model each cell provides a mean fraction $1-\phi = I_1/\log_2(S)$ of the total information required to discriminate S stimuli perfectly; any element of information has a chance (ϕ) to be missed by a cell; and being coded by one cell in a random relation to what is coded by other cells, it has a chance ϕ^C to be missed by C cells. $I^C = (1-\phi^C)\log_2(S)$ is therefore the information provided by C cells. As pointed out by Rolls, Treves and Tovée,[5] this model predicts zero redundancy in the limit of very large S, provided the information provided by single cells does not grow proportionately to the ceiling. This is because the overlap, or degree of redundancy, is $1-\phi$, which tends to zero for very large sets of stimuli. Thus the observed redundancy should not be taken as a limit to the information processing capabilities, but only to their performance to a given stimulus set. This hypothesis is supported by a study on the degree of redundancy in V1 neurons.[28] Increasing the number of stimuli more than halves the reported redundancy in these cells.

Failure to take into account ceiling effects, to select the appropriate information quantities and to measure them by correcting for limited sample size, easily results in artefactual findings, such as the postulated tendency of novel information to decrease with the inverse square root at the population size.[10] With only five stimuli in their set, Gochin et al[10] had a rather low ceiling on the information required to encode the set, and the failure to take this into account, together with the very few population sizes considered, makes their conclusions subject to artifact.

The 50 msec time sample highlights the fact that information can be made available more rapidly if the responses of a population of neurons are monitored rather than those of a single neuron. If the representation provided by each neuron has a low correlation, then the information will be available very rapidly, whatever subset of neurons is taken. This rapid availability of information from a population of neurons is one factor which contributes to the very rapid processing information from stage to stage in the visual cortical areas.

How Large Should Cell Populations Be?

Several factors have a bearing on the size of populations, driving it either up or down. Three factors favor large populations:

1. If one accepts that the information signaled by a population rises linearly with increasing cell numbers, and that contrary reports are due to ceiling effects, then larger populations will clearly be a more effective use of the limited number of neurons available. As the number of stimuli that can be represented by a population rises exponentially with increasing cell numbers, a large population will encode far more stimuli than two populations of the same total number of cells. In temporal visual cortex, it is possible that a population of 15 face-selective neurons could represent as many as 192 faces (at 50% accuracy), but a population of 30 neurons could represent up to 12,288 faces.[20]
2. If one assumes that the firing rate carries all the stimulus information, and that temporal coding is unimportant, then the question is how rapidly one can extract a reliable estimate of firing rate from the spike train. As we have seen in chapter 3, even looking at a single neuron, sampling the first 20-50 msec can provide an accurate prediction of the firing rate over 400 msec. If one then looks at populations of about 100 neurons, firing rate can be calculated from samples of just a few msec.[29-30] Thus, larger populations can potentially signal information more rapidly.
3. If the noise carried by a set of neurons is uncorrelated, then by averaging the signal between neurons it should be possible to filter out the noise to give a clear signal. The more neurons in this averaging process, the clearer the signal will be.

Thus, larger populations will have be able to represent more stimuli, more quickly and more accurately. However, limiting factors push the population size down. Neurons are grouped into columns and cortical areas which have common response properties and similar response profiles. Under such conditions, it is inevitable that there will be a degree of redundancy, if lower than that initially proposed by some authors.[9-10] This will limit the advantage of adding neurons to a population above a finite size. Additionally, the existence of common, correlated noise within cortical columns, which cannot be filtered out by pooling responses, also acts as a limiting factor in the number of neurons.[11]

These competing sets of factors limit the potential size of populations. Moreover, as factors such as response-tuning and redundancy seem to vary in different areas and at different processing levels, the size of populations is also likely to vary. However, most studies have settled on populations in the order of 100 neurons.[3,29,31] Given the exponential rise in the number of stimuli represented with the numbers of neurons in the population, 100 neurons could represent a vast number of stimuli, even with a modifier for redundancy and correlated noise.

References

1. Konorski J. Integrative activity of the brain: An interdisciplinary approach. Chicago 1967; University of Chicago Press.
2. Barlow HB. Single units and sensation: A neuron doctrine for perceptual psychology. Perception 1972;1:371-394.
3. Young MP, Yamane S. Sparse population coding of faces in the inferotemporal cortex. Science 1992; 256:1327-1331.
4. Rolls ET, Tovée MJ. Sparseness of the neuronal representation of stimuli in the primate temporal visual cortex. J Neurophysiol 1995; 73:713-726.
5. Rolls ET, Treves A, Tovée MJ, Panzeri S. Information in the neural representation of individual stimuli in the primate temporal visual cortex. J Comput Neurosci 1997; 4:309-333.
6. Rolls ET, Baylis GC, Hasselmo ME, Nalwa V. The effect of learning on the face selective responses of neurons in the cortex in the superior temporal sulcus of the monkey. Exp Brain Res 1989; 76:153-164.
7. Rolls ET, Tovée MJ, Ramachandran VS. Visual learning reflected in the responses of neurons in the temporal visual cortex of the macaque. Soc Neurosci Abstract 1993; 19:27.
8. Tovée MJ, Rolls ET, Ramachandran VS. Rapid visual learning in the neurons of the primate temporal visual cortex. Neuroreport 1996; 7:2757-2760.
9. Gawne TJ, Richmond BJ. How independent are the messages carried by adjacent inferior temporal cortical neurons? J Neurosci 1993; 13:2758-2771.
10. Gochin PM, Colombo M, Dorfman GA, Gerstein GL, Gross CG. Neural ensemble coding in inferior temporal cortex. J Neurophysiol 1994; 71:2325-2337.
11. Hurlbert AC, Derrington AM. How many neurons does it take to see? Current Biology 1993; 3:510-512.
12. Churchland PS, Ramachandran VS, Sejnowski TJ. A critique of pure vision. In: Koch C, Davis J eds. Large Scale Theories of The Brain. Cambridge Massachusetts: MIT Press. 1992.
13. Hinton GE, McClelland JL, Rumelhart DE. Distributed representations. In: Rumelhart DE, McClelland JL, eds. Parallel Distributed Processing, vol. 1, Cambridge: MIT Press 1996; 77-109.
14. Tovée MJ, Rolls ET. Information encoded in short firing rate epochs by single neurons in the primate temporal visual cortex. Visual Cognition 1995; 2:35-58.

15. Hasselmo ME, Rolls ET, Baylis OC.The role of expression and identity in the face-selective responses of neurons in the temporal visual cortex of monkey. Behav Brain Res 1989; 32:203-218.
16. Perrett DI, Hietnan JK, Oram MW, Benton PJ. Organization and functions of cells responsive to faces in the temporal cortex. Phil Trans R Soc Lond 1992; 8 335:23-30.
17. Georgopoulous AP, Lurito JT, Petrides M, Schwartz AB, Massey JT. Mental rotaion of the neuronal population vector. Science 1989; 243:234-236.
18. Georgopoulous AP, Schwartz AB, Kettner RE. Neuronal population of movement direction. Science 1986; 233 1416-1419.
19. Rolls ET, Treves A, Tovée MJ. The representational capacity of the distributed encoding of information provided by populations of neurons in the primate temporal visual cortex. Exp Brain Res 1997; 114:149-162.
20. Abbott LF, Rolls ET, Tovée MJ. Representational capacity of face coding in monkeys. Cereb Cortex 1996; 6:498-505.
21. Bialek W, Rieke F, de Ruyter van Steveninck RR, Warland D. Reading a neural code. Science 1991; 252:1854-1857.
22. Salinas E, Abbott LF. Vector reconstruction from firing rates. J Comput Neurosci 1994; 1:89-107.
23. Treves A. Graded response neurons and information encoding in autoassociative memories. Phys Rev A 1990; 42:2418-2430.
24. Treves A, Rolls ET. What determines the capacity of autoassociative memories in the brain. Network 1991; 2:371-397.
25. Treves A, Rolls ET. A computational analysis of the role of the hippocampus in memory. Hippocampus 1994; 4:374-391.
26. Barnes CA, McNaughton BL, Mizumori SJ, Kim LH. Comparison of spatial and temporal chracteristics of neuronal activity in sequential stages of hippocampal processing. Prog Brain Res 1990; 83:287-300.
27. Jung MW, McNaughton BW. Spatial selectivity of unit activity in hippocampal granular layer. Hippocampus 1993; 3:165-182.
28. Gawne TJ, Kjaer TW, Hertz JA, Richmond BJ. Adjacent visual cortical complex cells share about 20-percent of their stimulus related information. Cereb Cortex 1996; 6:482-489.
29. Oram MW, Perrett DI. Time course of neural responses discriminating different views of the face and head. J Neurophysiol 1992; 68:70-84.
30. Softky WR, Koch C. The highly irregular firing of cortical cells is inconsistent with temporal integration of random EPSPs. J Neurosci 1993; 13:334-350.
31. Shadlen MN, Newsome WT. Is there a signal in the noise? Curr Op Neurobiol 1995; 5:248-250.

CHAPTER 6

Visual Integration

The Problem of Integration

It is now widely accepted that the brain processes sensory information in a distributed, hierarchical system, although the precise organization of the hierarchies is uncertain at the present.[1-4] In these hierarchies, including the visual system, different features of the incoming sensory information, such as color, texture, shape, motion and so on, are processed separately and in parallel.[1-2] At some point in the visual hierarchy, the spatially segregated representations of the different visual features of a stimulus must be recombined to form a unified perception of the visual world. How the brain solves this "binding-problem" is a matter of some controversy. It has been proposed that neurons responding to the same object could be grouped into assemblies by invoking a temporal dimension to the responses of cells.[5-7] If the activity of the population of neurons responding to a stimulus were to synchronize, then the cells could be bound together into a "temporal assembly" (see Fig. 6.1). Temporal correlations of spikes would increase the saliency of neurons' responses by increasing the chance of simultaneous excitatory postsynaptic potentials (EPSPs)[8-9] and by allowing changes in synaptic efficiency to be evoked more easily.[10] Experimental evidence for this theory came when synchronous activity was found in, and between, various areas of the visual system in response to visual stimuli that would be expected to evoke feature binding.

In the cat visual system, a number of studies have shown that spatially separate cells within individual visual areas have shown synchronized activity, including the retina, the LGN, and areas 17, 18, 19 and the posteromedial lateral suprasylvian area (PMLS)[11-18] (Fig. 6.2). The majority of these studies have been in anesthetized animals, but this activity has also been shown to occur in awake cats as well.[19-22] However, this activity may not necessarily be linked to a

The Speed of Thought: Information Processing in the Cerebral Cortex,
by Martin J. Tovée.

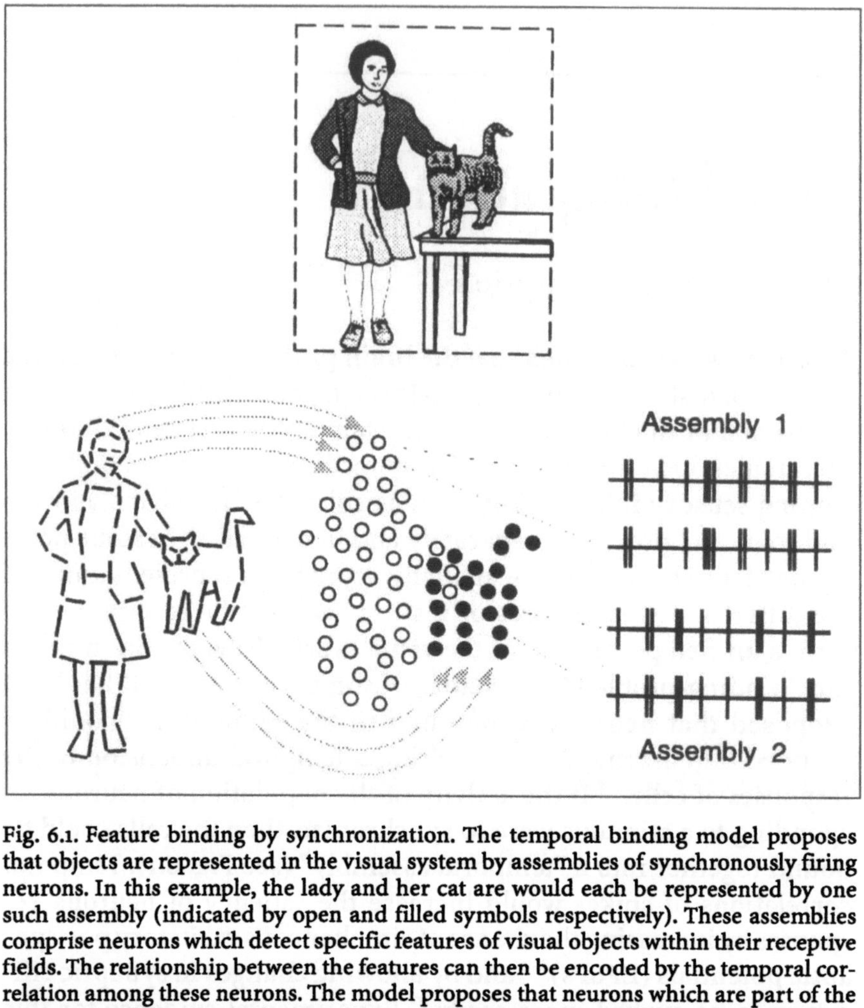

Fig. 6.1. Feature binding by synchronization. The temporal binding model proposes that objects are represented in the visual system by assemblies of synchronously firing neurons. In this example, the lady and her cat are would each be represented by one such assembly (indicated by open and filled symbols respectively). These assemblies comprise neurons which detect specific features of visual objects within their receptive fields. The relationship between the features can then be encoded by the temporal correlation among these neurons. The model proposes that neurons which are part of the same assembly fire in synchrony, whereas no consistent temporal relationship is found between cells belonging to different object representations. (Reproduced with permission from Engel AK, Roelfsema PR, Fries P, Brecht M, Singer W, Cereb Cortex 1997; 7: 571-582. © (1997) Oxford University Press).

temporal binding mechanism. Adjacent neurons with similar response properties would have a large degree of common input which would lead directly to synchronized or correlated activity. Consistent with this explanation is the fact that the degree of synchronization between neurons within a given area weakens with the distance between the recording electrodes and with increasing dissimilarity in the stimulus preferences of the recorded neurons.[23] Synchronized activity can also occur between neurons in different visual areas;

synchronization has been reported between neurons in the retina, LGN, 17, 18, 19 and PMLS.[13,22,24-25] Additionally, synchronization has been reported between neurons in equivalent areas in the two hemispheres.[26] But, again, a degree of correlated activity might be expected between areas within a processing stream.

The situation is less clear in primates. In the macaque V1, one study showed that, with very similar conditions of anesthesia and stimulation, the oscillations of the type reported in the cat were absent,[27] but oscillations have been reported in the macaque V1 by Eckhorn et al[28] and Friedman-Hill et al[29] and by Livingstone[30] in the squirrel monkey V1. Singer and his colleagues have reported oscillations in macaque V5,[31-32] but other groups found no such oscillations.[27,33] Nakamura and his colleagues[34] reported oscillations in macaque IT, but two other groups failed to find any evidence of oscillations in this area.[27-28] A degree of correlated firing has been reported between neurons of different primate visual areas,[36-38] and EEG studies have provided evidence for synchronization in the human visual system.[39-41]

Where's the Beef?

The existence of oscillatory activity is not proof that these oscillations function as part of a temporal binding mechanism. To provide evidence for this hypothesis, different research groups have made microelectrode recordings in the cat visual system to determine whether features of the oscillatory activity (such as their frequency and whether different populations of neurons synchronize) alter in a manner consistent with a role in temporal binding.

There are as many feedback connections between areas as there are feedforward connections. In cats, the feedback projections may play a role in synchronizing the responses of cells projecting to the primary visual cortex. Some LGN cells are reported to show a greater degree of response synchronization when stimulated by visual stimuli than would be expected if the responses of the cells were independent.[42] The synchronization is between cells activated by the same visual stimulus feature, in this case a moving light bar. If the same pair of cells is coactivated by two spatially segregated, simultaneous light spots, their responses do not synchronize. Moreover, this synchronization is abolished by the removal of the visual cortex, which suggests that it is generated by the thalamic feedback projection. Simple cells in V1 receive converging inputs from LGN cells to

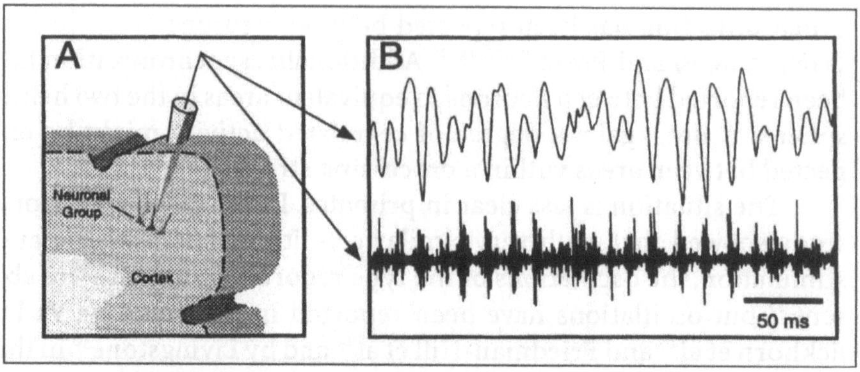

Fig. 6.2. Example of an oscillatory response in the cat V1. (A) in this experiment, the activity of a small group of neurons was recorded by means of a low-impedance electrode. (B) By filtering in different bandpasses, a local field potential (top trace) and multi-unit activity (bottom trace) can be extracted from the raw electrode signal. As a light bar is passed through the cell's receptive field, a clear oscillatory response is observed in the field potentials, indicating that the neurons of the recorded cluster have engaged in a coherent and rhythmic firing pattern. (Reproduced with permission from Engel AK, Roelfsema PR, Fries P, Brecht M, Singer W, Cereb Cortex 1997; 7: 571-582. © (1997) Oxford University Press).

produce their responses to lines and edges. Singer has suggested that synchronization is established when an array of LGN cells is activated by a contour whose orientation matches the preference of those cortical cells which receive convergent input from that array of LGN cells.[8] In short, it is suggested that feedback from V1 labels those LGN cells signaling a particular feature such as a line or an edge.

The temporal binding theory predicts that the phase relations between oscillating cells reflect the degree to which the cells are responding to the same object. The lack of overlap of the receptive fields of spatially separated cells has been exploited to test this important prediction. On two occasions, phase coherence between 2 recording sites with similar orientation preference was reported to be best when their receptive fields were stimulated by a single long bar.[14] It was less strong when the two receptive fields were stimulated by two short bars moved in the same direction, as would be consistent with a single bar moving behind an object. Oscillatory synchronization was absent when the two short bars were moved in opposite directions.[14] The reduction in phase coherence during presentation of a partially occluded bar compared to that during presentation of an unoccluded bar is surprising: This is just the situation when, according to the theory, a binding mechanism would be expected to operate, yet the proposed binding mechanism was weakened in this situation.

Another important prediction of the oscillation-binding theory was tested in a further report which concerned cells with overlapping receptive fields, but different orientation preferences.[43] A single light bar, whose orientation was midway between the optimal orientations for two sets of neurons, stimulated both sets of neurons and gave rise to phase coherent responses when swept across the neurons' receptive fields (Figs. 6.3 and 6.4). When two bars at the preferred orientations of the two neurons were moved across the receptive fields, both sets of cells were again stimulated but coherence was no longer apparent between cells with different orientation preferences. A more elaborate version of this experiment used four sets of neurons. The degree of coherence in the responses of neurons was dependent on the orientation of the stimulus bar. This experiment was interpreted as evidence for the cells moving into different ensembles by changing the temporal relationships of their firing patterns. Like the "long bar" experiment, above, this conclusion appears drawn from very few empirical examples: No other instance of four cells changing their relative temporal synchronization in this way has been reported in detail, and no further reports of the phenomena in the "long bar" experiment have appeared.

The methods that Singer's group used to analyze their data have been criticized, and it has been suggested that they may have overestimated the proportion of neurons displaying synchronized oscillatory activity (for a detailed critique see Young et al[27]). In the latter two experiments described above, the number of positive results is very low, and if the analysis technique is flawed, they may be false positives. Limited re-analysis of a subset of the data from the Singer's group, taking into account only some of these criticisms, demonstrates significant overestimation of the number of sites showing phase coherence.[44] Thus, the proof that synchronization of oscillations can act as a temporal binding mechanism is far from conclusive.

Neural Physiology

Synchronous neuronal activity is most easily observed in local field potentials (a measure of the activity of multiple cells near the electrode) and is less easy to detect in single-unit recordings, although

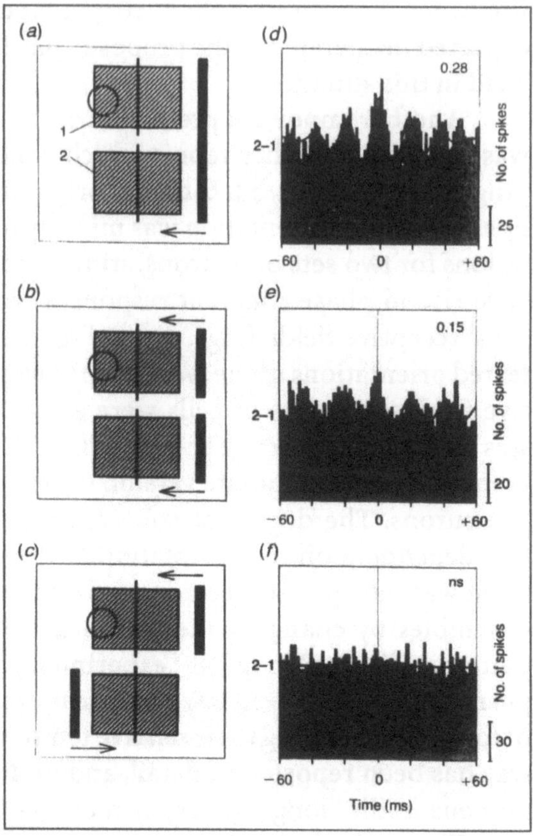

Fig. 6.3. An example of how the proposed synchronization of neural responses is influenced by visual stimuli. Multi-unit activity was recorded from two sites in cat V1 that were separated by 7 mm. The cells responded most strongly when the stimulus bar was at a vertical orientation. (a), (b), (c) show plots of the receptive fields (shown as hatched rectangles) under three different stimulus paradigms: (a) A long continuous light bar moving across both fields, (b) two independent light bars moving in the same direction, (c) the same two bars moving in opposite directions. The circle represents the center of the cat's visual field, and the black line drawn across each receptive field indicates the preferred orientation of those cells. (d), (e), (f) show correlograms obtained with each stimulus paradigm. Using the long light bar, the two oscillatory responses were synchronized, as indicated by the strong modulation of the cross-correlogram with alternating peaks and troughs (d). If the continuity of the stimulus was interrupted, the synchronization became weaker (e), and it totally disappeared if the motion of the stimulus was not coherent (f). The change in the stimulus configuration affected neither the strength or oscillatory nature of the two responses. The graph superimposed on each of the correlograms represents a Gabor function that was fitted to the data to assess the strength of the modulation. The number in the upper right corner indicates what Singer and his colleagues call the "relative modulation amplitude", a measure of the correlation strength that was determined by calculating the ratio of the amplitude of the Gabor function to its offset. Abbreviation: ns, not significant. Scale bars indicate the number of spikes. (Reproduced with permission from Engel AK, König P, Kreiter AK, Schillen TB, Singer W, Trends Neurosci 1992; 15:218-226. © (1992) Elsevier Trends Journals).

synchrony between neurons can be established by cross-correlation analysis of the recordings from pairs of neurons.[45] Earlier reports on binding tended to focus on neural oscillations (in the 30-100 Hz range) rather than synchrony,[7,13] but the actual function of oscillations in temporal binding is probably negligible. It is the synchronization of activity between neurons which is potentially important.[46]

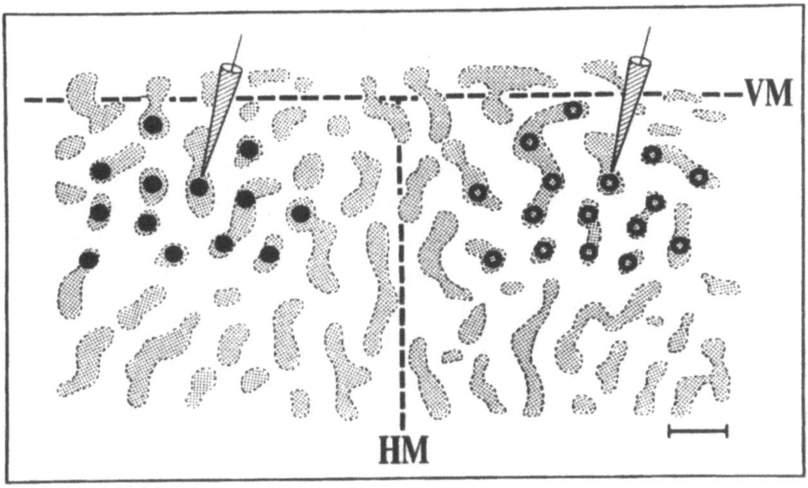

Fig. 6.4. A schematic of the long-bar experiment illustrated in Figure 6.3. The location of the recording electrodes is mapped onto a representation of the orientation columns in cat V1. Shaded areas correspond to bands of orientation columns responding to vertical contours. Electrode symbols indicate the two groups of cells from which recordings were taken. The diagram illustrates the measurements in 6.3(c) and (f), where recordings were taken in response to light bars moving in opposite directions. The two cell groups were assumed to be part of two large assemblies (represented by filled circles and filled circles), which are activated by two light bars. It is suggested that the members of each of the assemblies oscillate in synchrony, whereas no constant temporal relationship exists between them. The two assemblies are spatially separate. When the long continuous light bar is used, it is proposed that the intervening shaded orientation bands would also be activated, creating the formation of a single large assembly of synchronously oscillating groups. The scale bar is 1mm. Abbreviations: HM, representation of the horizontal meridian; VM, representation of the vertical meridian. (Reproduced with permission from Engel AK, König P, Kreiter AK, Schillen TB, Singer W, Trends Neurosci 1992; 15:218-226. © (1992) Elsevier Trends Journals).

Oscillatory activity appears to accompany synchrony in many cases, particularly in neighboring cells that receive a common input, and has been suggested to play a role in the establishment of synchrony.[47-48] Some simulation studies have suggested that the establishment of synchronization between neurons may be facilitated if these neurons show oscillatory firing patterns.[49-51] These models suggest that, based on the recurrent temporal structure of such patterns, reciprocally coupled neurons can entrain each other and improve synchrony within a few oscillatory bursts.[48] However, in reality the temporal constraints inherent in visual processing mean that this synchronization may be irrelevant to perception.

Human subjects can identify stimuli and bind features, such as color and shape, even when the stimuli are presented briefly and are masked afterwards so that the processing machinery is immediately reoccupied by another stimulus.[52-54] These masking experiments suggest that object recognition can occur when neurons are active for only 20-30 msec. Furthermore, analysis of the information signaled by neurons shows that the majority of information in a neural spike train is available within the first 50 msec and that less than 50 msec is sufficient for object perception.[55-56] This will allow for an average of only two spikes for a neuron firing at 40 Hz. Therefore, for synchronization to provide a binding mechanism under these conditions, it would have to occur very close to the stimulus onset. However, synchronization begins at a variable time after stimulus presentation and is not phase-locked to the onset of a stimulus.[45] It then seems unlikely that synchronization would play a crucial role in binding during everyday perception of familiar objects, although it might have a role in recognition learning, which may have a longer time course.[10]

If oscillatory activity is important for the establishment of synchrony, then oscillations which are not a response to a visual feature will seriously disrupt the system. Modeling studies have suggested that neurons may be inherent oscillators which can entrain neighboring cells into synchrony.[57] Such entrainment may account for many results, including those from experiments which show reduced synchronization in cats with induced strabismic amblyopia (squint)[58] (see Fig. 6.5). In strabismic cats the level of connectivity between different ocular dominance columns is drastically reduced,[58] so perhaps it is not surprising that synchrony is reduced.[46]

A further problem is whether neurons are actually capable of coincidence detection (see chapter 3). To preserve the temporal information in a synchronous assembly, the postsynaptic cells must respond on a millisecond time scale to EPSPs. However, it is more likely that cortical neurons may in fact be "integrate and fire" devices which summate or average EPSPs over a longer time scale of at least seven or more msec.[59-60] Such neurons would not be able to support the relatively high frequency code the temporal binding mechanism is postulated to use.

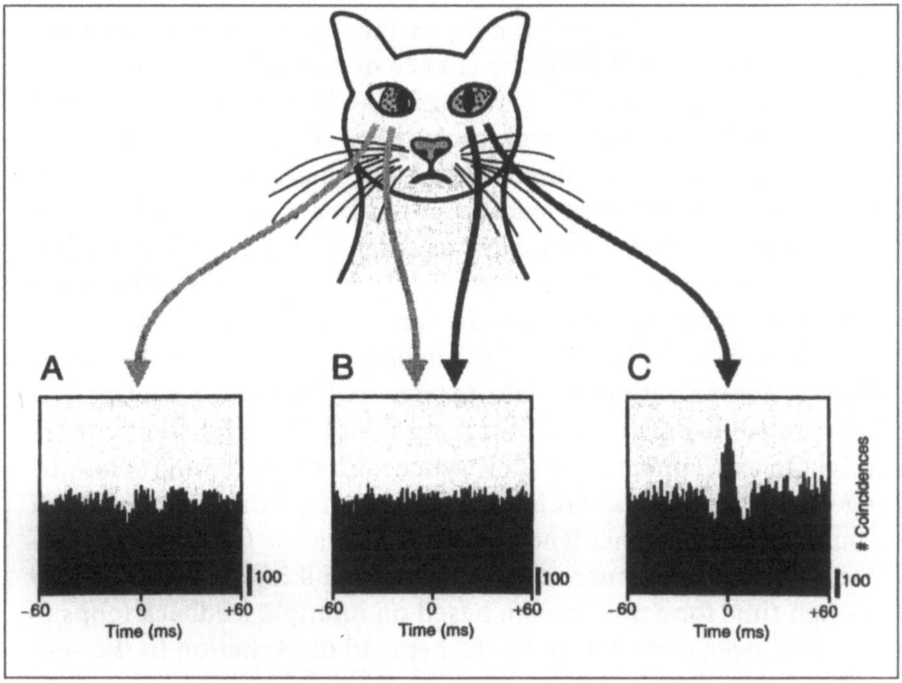

Fig. 6.5. Neuronal synchronization in V1 of strabismic amblyopia. The lower panel shows examples of cross-correlograms between cells driven by the normal eye, the strabismic eye and between cells dominated by different eyes. Temporal correlation is strong if both recording sites are driven by the normal eye (C). Synchronization is, on average, much weaker between cells dominated by the amblyopic eye (A) and in most cases negligible if the recording sites receive their input from different eyes (B). (Reproduced with permission from Engel AK, König P, Kreiter AK, Schillen TB, Singer W, Trends Neurosci 1992; 15:218-226. © (1992) Elsevier Trends Journals).

Coherence by Coincidence?

Synchrony which can be attributed purely to chance is normally excluded from correlograms by appropriate methods of data analysis. It is, however, still useful to realize how much synchronous activity exists in the system naturally, given the raised firing rates of cells due to stimulation. If the interval in which at least two inputs of a coincidence integrator cell must spike to make the cell fire is in the order of 1 msec,[45] then two independent inputs firing at 100 Hz will certainly set off the integrator at some moment within 100 msec (given random spike trains, the chance of coincident firing within any 1 msec interval is $0.1^2 = 0.01$, so the coincidence in 100 msec becomes about 1; similarly, the chance of coincident firing for two neurons discharging with an average of 40 Hz becomes 0.96 within 600 msec).[46] With such a high probability of chance "background" synchrony, it would be impossible for a single coincident pair of spikes to signal coherent features. A binding mechanism based on synchrony would therefore have to employ a more robust mechanism, for instance recurrent synchrony. However, this reduces the overall efficiency of binding to the same level as that of a rate code.

This could explain the failure to observe stimulus-locked onset of the synchronization; it could be triggered by a random event. The high probability of coincident spiking at high firing rates lends more weight to experiments that show synchronized processing at low firing rates.[61] How does continuous synchronized firing come about? Could an initial coincidence help, via lateral or feedback connections, to establish recurrent synchronized spiking? Is there actually enough time for a mechanism based on multiple feedback loops to be the answer, considering the fast spread of excitation in the ventral and dorsal streams? For example, there is a latency of less than 100 msec in the passage of activity from the retina to the highest association area (the anterior IT) in this stream.[62]

Alternative Solutions

If temporal synchrony does indeed prove to be an artifact or only a partial solution to the binding problem, then what other possibilities remain? One suggestion comes from Anne Triesman and her colleagues,[63-64] who propose that the spatial information from the dorsal stream, which extends from occipital to parietal cortex and is concerned with spatial locations within the visual field, may play a role. A neurological patient with bilateral lesions to occipital and parietal lobes has recently been described who appears to be impaired in binding the features of objects together to form a coherent percept.[63] This patient is simultagnosic (unable to attend to multiple objects), miscombines the colors of two simultaneously presented colored patches, makes frequent illusory-conjunctions between separate features and has great problems describing the location of objects in space. Triesman suggests that a spatial attention mechanism, which may be responsible for some aspects of feature binding in normal subjects, is damaged in this patient.[64] It would be surprising if the brain did not make use of the spatial information freely available to it to at least partially solve the binding problem.

Perhaps the simplest solution, however, may lie in the basic anatomy of the brain. It is often assumed that the binding problem cannot be solved by the processing of information in a "pyramidal" fashion, in which the information converges at a high level in the hierarchy, because the information that must be bound is too great to represented anatomically. However the actual amount of information the visual system extracted from the visual field is actually relatively low. For example, in primates high acuity is supported by only the central 2° of the retina, based on the fovea. The packing of the cone cells, which mediate both high acuity and color vision, declines rapidly with increasing eccentricity. The cone density decreases by a factor of 30 between the center of the fovea and an eccentricity of 10°.[65] The number of photoreceptors synapsing onto a single ganglion cell also increases with increasing eccentricity.[66] As a result, the "grain size" of the picture increases with increasing retinal eccentricity. This is also true of color, where color discrimination thresholds rise as an exponential function of eccentricity.[67] These responses are reflected in the relative proportions of neural space given over to processing the input from central and peripheral retina. The representation of the central visual field in the primary visual cortex is up to six times larger than that for the periphery.[68] The cumulative result of this organization is to create almost a tunnel vision effect, where only the visual information from the center of the visual field is fully sampled and analyzed[69] (Fig. 6.6). This is accentuated by the effects of attention. If a monkey's attention is focused on a stimulus, then the response of a visual cell to that stimulus is not enhanced, but the cell's response to other stimuli within its receptive field is reduced.[70] The net effect of attention seems to be to reduce the functional size of a cell's receptive field.

Anatomical studies suggest that there is a considerable reconvergence of information in the frontal lobe, rostral superior temporal sulcus and the limbic system.[71-73] The high information capacity of distributed neural populations (see chapter 5), combined with the comparatively low information input coming into the visual system, suggests that coherent representations of objects could plausibly be brought about by physical convergence of the information at higher levels of the system.

Fig. 6.6. On the left is a photograph of Malcolm Young and his students in the laboratory. The visual system provides us with the sensation that all aspects of the scene are simultaneously processed. On the right is a processed version of the same photograph, giving an impression of what the retina of a viewer might signal to the brain about this scene. Only a restricted region corresponding to the central few degrees of vision is fully processed; our impression of a complete scene is built up from a number of "snap shots" made by the constantly moving eye. (Reproduced with permission from Young MP, Curr Biol 1993; 4: 51-53. © (1993) Current Biology)

Other Roles for Synchrony

It has recently been shown that the neural responses may synchronize without a rise in neuronal firing rate.[62] It has been suggested that auditory cortical neurons may encode the features of a stimulus, rather than group features together, by coordinating their action potentials without increasing their firing rate. Coincident firing by primary cortical neurons will obviously increase the firing rate of their higher-level target cells. Thus, coordination of responses could encode stimulus features in primary sensory areas, while evoking an explicit representation of the object features at higher levels.[46] A code which does not require a rise in firing rate is appealing, as it allows coding to occur under various conditions, such as with high stimulation and after habituation, thus adding another dimension to the coding achievable by neurons.

Evidently, if synchrony is the key to object representation, then the situation is far more complicated than previously assumed. As appealing as this more subtle version of temporal binding is, it still falls foul of the incredible speed of perception.[63] The establishment of coincidence or synchrony is simply too slow to account for normal object perception. The most obvious alternative role for synchrony is none at all. Synchrony may well just be an artifact caused by common input to the cells which show synchronous activity.[46,60,74]

References

1. Zeki S, Shipp S. The functional logic of cortical connections. Nature 1988; 335:311-317.
2. Van Essen DC, Gallant JL. Neural mechanisms of form and motion processing in the primate visual system. Neuron 1994; 13:1-10.
3. Hilgetag CC, O'Neill MA, Young MP. Indeterminate organization of the visual system. Science 1996; 271:776-777.
4. Scannell JW, Blakemore C, Young MP. Analysis of connectivity in the cat cerebral cortex. J Neurosci 1995; 15:1463-1483.
5. Milner PM. A model for visual shape recognition. Psychol Rev 1974; 81:521-535.
6. von der Malsburg C. The correlation theory of brain function. Internal Report 1981; 81-2 Grottengen: Max-Planck-Institute for Biophysical Chemistry. Reprinted In: Models of neural networks II, Domany E, van Hemmen JL, Schulten K eds. Berlin: Springer 1994; 95-119.
7. von der Malsburg C, Schneider W. A neural cocktail-party processor. Biol Cybern 1986; 54:29-40.
8. Singer W. Synchronization of cortical activity and its putative role in information processing. Ann Rev Physiol 1993; 55:49-374.
9. von der Malsburg C. Binding in models of perception and brain function. Curr Opin Neurobiol 1995; 5:520-526.
10. Ritz R, Sejnowski TJ. Synchronous oscillatory activity in sensory systems: New vistas on mechanisms. Curr Opin Neurobiol 1997; 7:536-546.
11. Toyama K, Kimura M, Tanaka K. Cross-correlation analysis of interneuronal connectivity in cat visual cortex. J Neurophysiol 1981; 46:191-201.
12. Tso's DY, Gilbert CD, Wiesel TN. Relationships between horizontal interactions and functional architecture in cat striate cortex as revealed by cross-correlation analysis. J Neurosci 1986; 6:1160-1170.
13. Eckhorn R, Bauer R, Jordon W, Brosch M, Kruse W, Munk M, Reitboeck HJ. Coherent oscillations: A mechanism for feature linking in the visual cortex? Biol Cybern 1988; 60:121-130.
14. Gray CM, König P, Engel AK, Singer W. Oscillatory responses in cat visual cortex exhibit intercolumnar synchronization which reflects global stimulus properties. Nature 1989; 338:334-337.
15. Gray CM, Singer W. Stimulus-specific neuronal oscillations in orientation columns of cat visual cortex. Proc Natl Acad Sci USA 1989; 86:1698-1702.

16. Engel AK, König P, Gray CM, Singer W. Stimulus-dependent neuronal oscillations in cat visual cortex: Intercolumnar interaction as determined by crosscorrelation analysis. Eur J Neurosci 1990; 2:588-606.
17. Schwartz C, Bolz J. Functional specificity of a long-range horizontal connection in cat visual cortex: A cross-correlation study. J Neurosci 1991; 11:2995-3007.
18. Brosch M, Bauer R, Eckhorn R. Synchronous high-frequency oscillations in cat area 18. Eur J Neurosci 1995; 7:86-95.
19. Chiang C, von Stein A, König P. Synchronous activity between primary visual and sensorimotor cortex in the awake behaving cat. Soc Neurosci Abstr 1996; 22:643.
20. Fries P, Roelfsema PR, Engel AK, König P, Singer W. Synchronized gamma frequency oscillations correlate with perception during binocular rivalry in awake squinting cats. Soc Neurosci Abstr 1996; 22:282.
21. Gray CM, Viana Di, Prisco G. Stimulus-dependent neuronal oscillations and local synchronization in striate cortex of the alert cat. J Neurosci 1997; 17:3239-3253.
22. Roelfsema PR, Engel AK, König P, Singer W. Visuomotor integration in association with zero time-lag synchronization among cortical areas. Nature 1997; 385:157-161.
23. Gray CM, Engel AK, Konig P, Singer W. Temporal properties of synchronous oscillatory neuronal interactions in cat striate cortex. In: Schuster HG ed. Nonlinear Dynamics and Neuronal Networks. Verlag 1991; 27-56.
24. Engel AK, Kreiter AK, König P, Singer W. Synchronization of oscillatory neuronal responses between striate and extrastriate visual cortical areas of the cat. Proc Natl Acad Sci USA 1991; 88:6048-6052.
25. Nelson JI, Salin PA, Munk MHJ, Arzi M, Bullier J. Spatial and temporal coherence in corticocortical connections: A cross-correlation study in areas 1 to 18 in the cat. Vis Neurosci 1992; 9:21-37.
26. Engel AK, König P, Kreiter AK, Singer W. Interhemispheric synchronization of oscillatory neuronal responses in cat visual cortex. Science 1991; 252:1177-1179.
27. Young MP, Tanaka K, Yamane S. On oscillating neuronal responses in the visual system of the monkey. J Neurophysiol 1992; 67:1464-1474.
28. Eckhorn R, Frein A, Bauer R,Wolbern T, Kehr H. High-frequency (60-90 Hz) oscillations in primary visual cortex of awake monkey. NeuroReport 1993; 4:243-246.
29. Friedman-Hill SR, Maldonado PE, Gray CM. Short and long-range synchronization in macaque striate cortex. Soc Neurosci Abstr 1995; 21:1505.
30. Livingstone MS. Oscilatory firing and interneuronal correlations in squirrel monkey striate cortex. J Neurophysiol 1996; 75:2467-2485.
31. Kreiter AK, Singer W. Oscillatory neuronal responses in the visual cortex of the awake macaque monkey. Eur J Neurosci 1992; 4:369-375.

32. Kreiter AK, Singer W. Stimulus-dependent synchronization of neuronal responses in the visual cortex of awake macaque monkey. J Neurosci 1996; 16:2381-2396.
33. Bair W, Koch C, Newsome W, Britten K. Power spectrum analysis of bursting cells in area MT in the behaving monkey. J Neurosci 1994; 14:2870-2892.
34. Nakamura K, Mikami A, Kubota K. Oscillatory activity related to short-term visual memory in monkey temporal pole. NeuroReport 1993; 3:117-120.
35. Tovée MJ, Rolls ET. Oscillatory activity is not evident in the primate temporal visual cortex with static stimuli. NeuroReport 1992; 3:369-372.
36. Roe AW, Ts'o DY. Functional connectivity between V1 and V2 in the primate. Soc Neurosci Abstr 1992; 18:11.
37. Frien A, Eckhorn R, Bauer R, Woelbern T, Kehr H. Stimulus-specific fast oscillations at zero phase between visual areas V1 and V2 of awake monkey. NeuroReport 1994; 5:2273-2277.
38. Salin PA, Bullier J. Corticocortical connections in the visual system: Structure and function. Physiol Rev 1995; 75:107-154.
39. Bressler SL, Coppola R, Nakamura R. Episodic multiregional cortical coherence at multiple frequencies during visual task performance. Nature 1993; 366:153-156.
40. Tallon C, Bertrand O, Bouchet P, Pernier J. Gamma-range activity evoked by coherent visual stimuli in humans. Eur J Neurosci 1995; 7:1285-1291.
41. Tallon-Baudry C, Bertrand O, Delpuech C, Pernier J. Stimulus specificity of phase-locked and nonphase-locked 40 Hz visual responses in human. J Neurosci 1996; 16:4240-4249.
42. Sillito AM, Jones EH, Gerstein GL, West DC. Feature-linked synchronization of thalamic relay cell firing induced by feedback from the visual cortex. Nature 1994; 369:479-482.
43. Engel AK, König P, Singer W. Direct physiological evidence for scene segmentation by temporal coding. Proc Natl Acad Sci USA 1991; 88:9136-9140.
44. Gray CM, Engel AK, Konig P, Singer W. Synchronization of oscillatory neuronal responses in cat striate cortex: Temporal properties. Vis Neurosci 1992; 8:337-347.
45. Singer W, Gray CM. Visual feature integration and the temporal correlation hypothesis. Ann Rev Neurosci 1995; 18:555-586.
46. Golledge HDR, Hilgetag C, Tovée MJ. Information processing: A solution to the binding problem? Current Biology 1996; 6:1092-1095.
47. König P, Engel AK. Correlated firing in sensory-motor systems. Curr Opin Neurobiol 1995; 5:511-519.
48. Engel AK, Roelfsema PR, Fries P, Brecht M, Singer W. Role of the temporal domain for response selection and perceptual binding. Cereb Cortex 1997; 7:571-582.

49. Schuster HG, Wagner P. Mutual entrainment of two limit cycle oscillators with time delayed coupling. Prog Theor Phys 1990; 81:939-945.
50. König P, Schillen TB. Stimulus-dependent assembly formation of oscillatory responses: I Synchronization. Neural Computat 1991; 3:155-156.
51. Sompolinsky H, Golomb D, Kleinfield D. Cooperative dynamics in visual processing Phys Rev A 1991; 43:6990-7011.
52. Rolls ET, Tovée MJ. Processing speed in the cerebral cortex and the neurophysiology of backward masking. Proc R Soc Lond B 1994; 257:9-15.
53. Rolls ET, Tovée MJ, Purcell DG, Stewart AL, Azzopardi P. The responses of neurons in the temporal cortex of primates and face identification and detection. Exp Brain Res 1994; 101:473-484.
54. Tovée MJ, Rolls ET, Panzeri S. A quantitative information analysis of the neurophysiology of backward visual masking. Eur J Neurosci 1998; submitted.
55. Tovée MJ, Rolls ET, Treves A, Bellis RP. Information encoding and the responses of single neurons in the primate temporal visual cortex. J Neurophysiol 1993; 70:640-654.
56. Tovée MJ, Rolls ET. Information encoded in short firing rate epochs by single neurons in the primate temporal visual cortex. Visual Cognition 1995; 2:35-58.
57. Cairns DE, Baddeley RJ, Smith LS. Constraints on synchronizing oscillator networks. Neural Computation 1993; 5:260-266.
58. Roelfsema PR, König P, Engel AK, Sireteanu R, Singer W. Reduced synchronization in the visual cortex of cats with strabismic amblyopia. Eur J Neurosci 1994; 6:1645-1655.
59. Shadlen MN, Newsome WT. Noise neural codes and cortical organization? Curr Opin Neurobiol 1994; 4:569-579.
60. Shadlen MN, Newsome WT. Is there a signal in the noise? Curr Opin Neurobiol 1995; 5:248-250.
61. de Charms RC, Merzenich MM. Primary cortical representation of sounds by the coordination of action potential timing. Nature 1996; 381:610-613.
62. Tovée MJ. How fast is the speed of thought? Current Biology 1994; 4:1125-1127.
63. Friedman-Hill SR, Robertson LC, Triesman A. Parietal contributions to visual feature binding: Evidence from a patient with bilateral lesions. Science 1995; 269:853-855.
64. Treisman A. A binding problem. Curr Opin Neurobiol 1996; 6:171-178.
65. Curcio CA, Allen KA, Sloan KR, Lerea CL, Hurley JB, Klock IB, Milam AH. Distribution and morphology of human cone photoreceptors stained with anti-blue opsin. J Comp Neurol 1991; 312: 610-624.
66. Perry VH, Cowey A. The ganglion cell and cone distributions in the monkey's retina: Implications for central magnification factors. Vision Res 1985; 25:1795-1810.

67. Nagy AL, Wolf S. Red-green discrimination in peripheral vision. Vision Res 1993; 33:235-242.
68. Azzopardi P, Cowey A. Preferential representation of the fovea in the primary visual cortex. Nature 1993; 361:719-721.
69. Tovée MJ. An Introduction To The Visual System. Cambridge University Press 1996.
70. Moran J, Desimone R. Selective attention gates visual processing in the striate cortex. Science 1985; 229:782-784.
71. Van Essen DC, Anderson CH, Felleman DJ. Information processing in the primate visual system: An integrated systems perspective. Science 1992; 255:419-423.
72. Young MP. Objective analysis of the topological organization of the primate cortical visual system. Nature 1992; 358:152-154.
73. Young MP, Scannell JW, O'Neill MA, Hilgetag CG, Burns G, Blakemore C. Nonmetric multidimensional scaling in the analysis of neuroanatomical connection data and the organization of the primate cortical visual system. Phil Trans R Soc Lond B 1995; 348:281-308.
74. Ghose GM, Freeman RD.Oscillatory discharge in the visual system: Does it have a functional rate? J Neurophysiol 1992; 68:1558-1574.

CHAPTER 7

The Speed of Thought

How Fast Can You See?

This book has explored how visual information is processed. An intrinsic constraint on this processing is temporal factors. As we have seen, a number of pieces of evidence point to very rapid processing. Analysis of the neural responses of face-selective cells in the temporal visual cortex using Bayesian information theory suggests that around 87% of the information available in a 400 msec sample is available from a 50 msec sample near the start of a spike train, and around 67% of the information is available in a 20-30 msec sample.[1-2] These results are consistent with other studies, such as that of Thorpe and Imbert,[3] who recorded from orientation selective neurons in the macaque V1 and found that orientation tuning curves appear very similar whether they are calculated from the firing rate based on a 50 msec sample or based on a 300 msec sample. The results of such studies suggest that most of the information encoded in a spike train is available in a short period at the beginning of the spike train.

An alternative approach is to examine how the whole system functions, by placing constraints on how long it has to process a visual stimulus. This can be done using a backward masking paradigm. Even when a mask stimulus follows the onset of the test stimulus at very short time intervals, pattern and object discrimination can still occur. For example, using face stimuli, at an interval between the test and mask stimulus of only 20 msec, human subjects can still just recognize and discriminate face identity. Under these conditions using the same experimental apparatus, single cell recording from rhesus monkeys shows that some face-selective cells in the temporal cortex fire for only around 20-30 msec.[4] The mask stimulus has the effect of strongly inhibiting the cell's response, presumably by lateral inhibition. These results suggest that the minimum

The Speed of Thought: Information Processing in the Cerebral Cortex,
by Martin J. Tovée.

amount of time neurons at each synapse which need to be active in order to mediate recognition and discrimination is in the region of 20-30 msec.

From Neuron to Neuron

The results discussed above are complemented by the findings from response latencies in different visual areas. As one moves along the visual pathways from the retina, through primary visual cortex to the visual association areas, the response characteristics of neurons change. Neurons higher up the pathways have larger receptive fields, respond to more complex stimuli and have greater response latencies. The first two characteristics are due to processing and integration of visual information in the preceding areas. The increasing response latencies represent the time to transmit the information through the brain and presumably the time taken to carry out some degree of processing at each stage. Visual neurons are usually selective for specific stimuli, and seem to show this specificity even at the very beginning of their response,[5] suggesting that the cells at the previous stage are carrying out some degree of processing before passing information on to the next stage.

Thus, one way to look at how long it takes to process information at a synapse is to look at the differences in the latency of response in consecutive visual areas. If one assumes that each synapse is processing information before passing it on, and not just acting as a relay station, and one knows how the visual system is wired up, one can then determine the delay introduced by the processing of visual information through an area of the cortex by comparing the response latencies of neurons in different areas. Unfortunately, although a great many studies have looked at the responses of single neurons in individual areas, few of them specifically address the issue of response latencies, and those from which such information can be derived display important methodological differences. Therefore, the most reliable measure of differences in response latencies come from groups who have used the same or comparable stimuli and experimental conditions to examine visual latencies in several different visual areas and sub-areas.

One group has recorded in macaque V1, V4 and V5 (or MT).[6-7] In common with previous studies in primates, they found that the cell responses fell into two main groups: early, transient responses and later, sustained responses. The former cell responses are thought to correspond to the dorsal pathway, and the latter responses to the ventral pathway, and this hypothesis has been supported by lesion experiments.[7] It is also consistent with the analysis of cell responses in the primate temporal cortex, where some degree of integration between form and motion seems to occur. In these cases it seems that information relating to motion and spatial position precedes the arrival of that on form.[8] The two pathways have different latencies, but comparable differences in latency between different areas. For example, the transient responses in V1 have an average latency of 28 msec and those in V5 have an average latency of 39 msec, a difference of 11 msec. The sustained responses have an average latency of 38 msec in V1, and an average latency of 50 msec in V4, a difference of 12 msec. A second group has also measured response latencies in these areas and in V2, and has found comparable differences in latency. For example, they report a difference of 11 msec between V1 and V2 and a difference of 9 msec between V1 and V5.[9-10] They also report a difference in latencies of 40 msec between cells in V1 and IT.[11] Given that between V1 and IT there are at least 2-3 synapses, then these findings yield a consistent synaptic latency of 10-15 msec in transmission from one area to another. This lag also seems to be common to the passage of information between different subdivisions of a given area. For example, there seems to be a lag of 15 msec between neurons in layer 4C and in the more superficial layers of V1, and a lag of 11 msec between viewer-centered cells and object-centered cells in the temporal visual cortex.[7,12] These results from primates are comparable to results from the cat visual system, which show delays of 5-15 msec between different visual areas.[13]

All the studies mentioned above report a considerable overlap in cell activity in different visual areas, such that most of the neurons at different stages in the visual system are simultaneously active.[13] It seems that a neuron is continually passing on information as it is processing it, rather than completing the processing and then passing it on. Over a period of time, different factors will influence the processing of information at a synapse. Initially it will be based purely on incoming information from the preceding areas (feedforward information), but at varying temporal intervals different feedback mechanisms will play a modulating role. Initially, this would be in the form of lateral inhibition, followed by intracortical feedback and then feedback from higher areas. This would be consistent with what is known of human reaction time studies. Simple detection of stimuli can be performed within 200 msec, whereas the recognition and discrimination of patterns leads to a lengthening of reaction times to 400-500 msec. This suggests that the longer an individual cell is active, the more sophisticated the information analysis will be. However, even with a reaction time of 400-500 msec for the recognition and discrimination of visual stimuli, there will still only be about 20-30 msec processing time per synapse.

Quick Thinking

We know that it is possible for humans to recognize and respond to a visual stimulus within 400-500 msec. If one considers that at least half of this time is involved in the generation and implementation of motor commands, then the processing and recognition of visual stimuli takes somewhat less than 200 msec. This is consistent with a field potential study which recorded from strips of stainless steel electrodes resting on the surface of the extrastriate cortex of epileptic patients being evaluated for surgery.[14] They reported a large amplitude negative potential generated bilaterally in regions of the mid-fusiform gyrus and inferior temporal gyri which was selective for faces. This response had an average latency of 200 msec, but could be as short as 140 msec. This finding is supported by an event-related potential (ERP) study using a go/no-go categorization paradigm in which subjects had to decide whether a previously unseen photograph contained an animal.[15] The study reported a negative potential specific to the no-go trials which developed roughly 150 msec after stimulus onset, suggesting that the visual processing needed to perform this task can be achieved in under 150 msec. These studies all suggest that information can be processed in the visual system within 150-200 msec. If one divides this time by the number of synapses in the visual pathways, it suggests rapid processing in the order of tens of milliseconds at each synapse.

Second Thoughts?

Such fast processing seems to suggest that visual processing, at least initially, is based purely on feedforward inputs, with little scope for modification by feedback input from lateral connections or from connections from subsequent processing areas. This assumes that the modification of a neuron's firing rate by lateral connections is intrinsically a feedback effect, for which the minimal requirement is that a cell contributes to the firing of another cell, and then receives back a spike from that cell and fires again, taking two interspike intervals (one each at the current rate of firing of each of the two cells). The process is then repeated a few times in order to lead to a steady activity distribution, which could possibly take up to several hundred milliseconds. However, this approach neglects the degree of spontaneous activity in the system.[16] The lateral connections may contribute much of their effect before genuine feedback occurs, even before most cells have had the time to produce a single spike in response to a new stimulus. For example, a spike emitted by a cell may immediately influence the firing of a second cell which is close to threshold, while more time is needed for the firing of this second cell to have an effect on the subsequent output of the first cell. One must also consider the fact that the influence, reciprocated or not, of a given cell onto another is of comparatively little importance for the collective behavior of the whole cell population, which is determined by thousands of inputs onto each cell.

Thus, in a large population of cells, firing at a low but significant rate, feedforward input will find some cells that are already close to their response threshold. The slight modification in the firing times of those cells (brought forward or postponed depending on whether each cell is excited or inhibited by the afferents) will be rapidly transmitted through lateral connections, and contribute to initiating a self-amplifying process of adjusting the firing pattern that would otherwise have been determined by external inputs alone.[16] The speed of the process will be determined by the time course of synaptic action, which (assuming transmission delays are close to negligible) means that the determining factor is the time constant of synaptic conductance. Therefore, information-rich neuronal responses emerging at relatively early stages may reflect not only feedforward processes, but also lateral interactions. These lateral interactions may or may not involve the propagation of spikes around feedback loops. The sequential activation of most such loops will require more than 20 msec, but the population response will already within the time be dependent on the activity of recurrent connections. The ability of recurrent connections to contribute usefully to information processing is due to the fact that their effects occur within a time determined primarily by the time constants of the synaptic conductances. Over the longer time scale of 50 msec or more, long range feedback loops, such as those based on backprojections from higher processing levels, are likely to be important for information processing. The importance of backprojections over the longer term has been demonstrated by the reversible cooling of V1 while recording from cells in the LGN.[17] The information signaled by the LGN cells, whether calculated from firing rate or from the temporal pattern of spike distribution, showed a significant reduction when V1 was cooled.

Rapid Visual Learning

Not only can the visual system process information rapidly, it can also rapidly update and modify the response properties of neurons on the basis of experience to optimize recognition. For example, people can accurately reconstruct meaningful objects out of fragmentary or ambiguous evidence.[18] However, naive subjects often experience a long latency before the object is recognized unless they have been previously exposed to an unambiguous version of the object in the appropriate context. Once they have recognized the unambiguous version of the object, even the ambiguous version of the stimulus can be recognized almost instantly. This represents a striking example of rapid perceptual learning, which is presumably implemented by modifying the responses of neurons in the visual system.

To explore the role of the face-selective neurons in visual learning, a recent study compared the responses of 18 human subjects to the responses of 21 face-selective neurons in the same visual learning paradigm.[19] In this paradigm, there were initially ten 0.5 msec presentations of an ambiguous image of a novel face (see Fig. 7.1), interleaved with 10 presentations of a control image (trial block 1). This was followed by 10 presentations of the same face, but in its unambiguous (gray scale) form interleaved with control images (trial block 2). There were then 10 presentations of the original ambiguous face used in trial block 1 and the control images (trial block 3). The control images were either ambiguous and unambiguous image pairs of complex objects, or ambiguous face images presented without being preceded by their unambiguous counterparts. Under these conditions, the human subjects were unable to recognize the ambiguous stimuli in trial block 1, but were able to recognize the unambiguous images in trial block 2 and the ambiguous images in trial block 3. This clearly demonstrated visual learning. At the neural level, the cells responded weakly to the ambiguous face in trial block 1, strongly to the unambiguous face in trial block 2 and then again strongly to the ambiguous face in trial block 3 (see Figs. 7.2 and 7.3).[19]

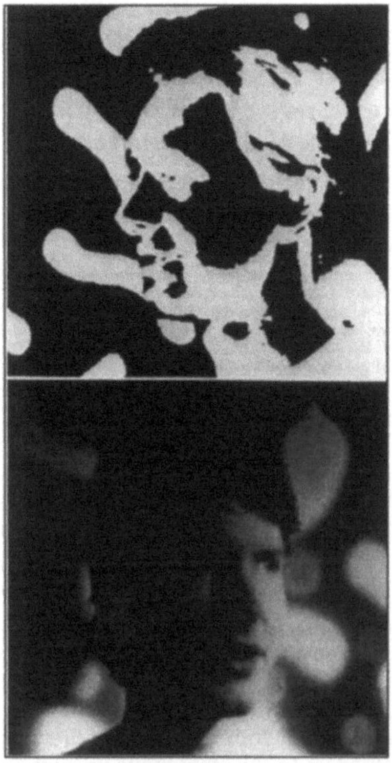

Fig. 7.1. Examples of the ambiguous (top) and unambiguous (bottom) face stimuli. The stimuli used were gray scale images and were presented on a VDU which subtended 8.5° of visual field. Each presentation of an image lasted for 0.5 sec. The unambiguous images were gray scale pictures of faces or of complex objects taken against a patterned background. The ambiguous images were binarized black and white versions of the unambiguous pictures which tended to camouflage the faces or objects so that could not easily be segmented from the background. These images resembled Mooney face images. (Reproduced with permission from Tovée MJ, Rolls ET, Ramachandran VS., NeuroReport 1996; 7: 2757-2760. © (1996) Rapid Science Publishers Ltd.)

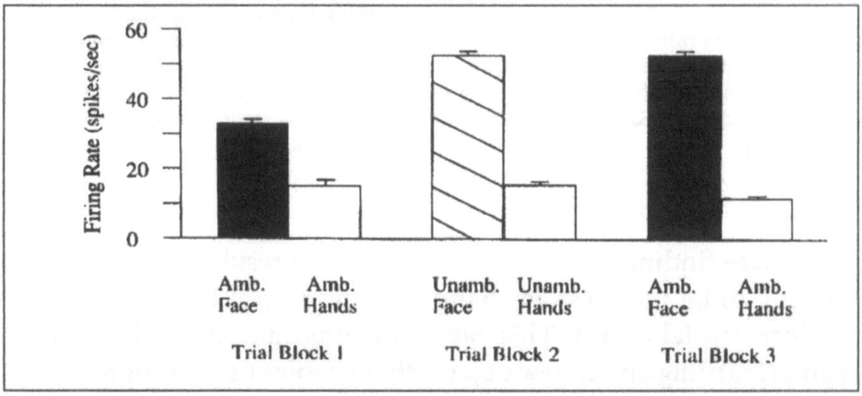

Fig. 7.2. For the first set of 10 presentations of each stimulus, the ambiguous face elicited a firing rate of 33 spikes/sec (averaged across all cases over the 10 trials in a block), and the control image, which was an ambiguous image of hands, elicited a firing rate of 15 spikes/sec. In stimulus block 2, the unambiguous version of the same face elicited a firing rate of 53 spikes/sec and the unambiguous version of the control elicited a firing rate of 16 spikes/sec. In stimulus block 3, the ambiguous stimulus elicited a significantly higher response (53 spikes/sec) than it had in block 1, but there was no significant increase in firing rate to the control stimulus (13 spikes/sec). This shows that there was a specific enhancement of the cell's response to the face in the ambiguous condition, not just a general increase in responsiveness to all images. (Reproduced with permission from Tovée MJ, Rolls ET, Ramachandran VS., NeuroReport 1996; 7: 2757-2760. © (1996) Rapid Science Publishers Ltd.)

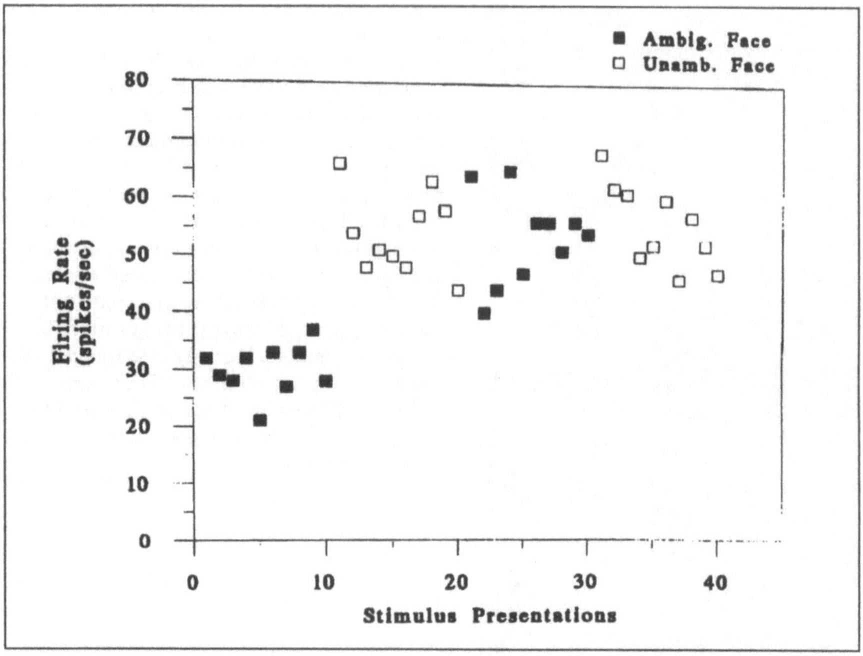

Fig. 7.3. The responses of the same cell as in Figure 7.2 to the three sets of stimulus presentation. Each square represents the responses of the neurons to a single stimulus presentation. The black squares represent the responses to the ambiguous version of the face image and the white squares to the unambiguous version of the image. The change in firing rate did not represent a gradual drift upwards in the strength of response over the course of the second ten presentations, but a sharp and significant change starting at the first response. (Reproduced with permission from Tovée MJ, Rolls ET, Ramachandran VS., NeuroReport 1996; 7: 2757-2760. © (1996) Rapid Science Publishers Ltd.)

These findings provide evidence for the results of rapid visual learning to be manifest as changes in the neural responses in the temporal visual cortex. This rapid learning may underlie the perceptual learning about new objects that enables us to recognize new objects or faces which have previously been seen for only a second or two. This rapid learning is what would be expected from neurons which are part of a system for learning about the rapidly changing views of objects which would enable us to recognize these objects subsequently. This rapid learning may also be important in learning other types of invariance, such as size and translation invariance.

The Speed of Thought

In conclusion, it seems that the visual system is capable of processing information very rapidly. This rapid processing of visual information is facilitated by the continuous passage of information from one stage to the next. This allows information to be processed in parallel at different stages in the visual pathways, offset from the preceding stage by 10-15 msec. Based on the activity of single neurons, most of the information seems to be encoded in the first 50-100 msec of neuronal activity, and only 20-30 msec activity is sufficient for the identification and discrimination of even complex visual stimuli. This rapid processing is enhanced by population encoding, which allows a unique pattern of activity to emerge much faster. This is just one of a number of factors which act as an upward pressure on population size.

This rapid processing seen in the visual system is likely to be duplicated in the other sensory systems and in the cerebral cortex as a whole. It is no good being able to rapidly process sensory information if one then does not have the cognitive mechanisms to rapidly assess this information and the motor systems to implement a response on a meaningful time scale. This rapid processing has important implications, both for the neurophysiologist and the time over which he or she measures a neuron's response and for the neural network modeller whose models must function in a biologically plausible time frame.

References

1. Tovée MJ, Rolls ET, Treves A, Bellis RP. Information encoding and the responses of single neurons in the primate temporal visual cortex. J Neurophysiol 1993; 70:640-654.
2. Tovée MJ, Rolls ET. Information encoded in short firing rate epochs by single neurons in the primate temporal visual cortex. Visual Cognition 1995; 2:35-58.
3. Thorpe SJ, Imbert M. Biological constraints on connectionist modelling. In: Pfeifer R, Schreter Z, Fogelman-Soulie F, eds. Connectionism in Perspective. Elsevier, 1989; 63-92.
4. Rolls ET, Tovée MJ. Processing speed in the cerebral cortex and the neurophysiology of backward masking. Proc R Soc Lond B 1994; 257:9-15.
5. Oram MW, Perrett DI. Time course of neural responses discriminating different views of the face and head. J Neurophysiol 1992; 68:70-84.

6. Maunsell JHR. Physiological evidence for two visual subsystems. In: Vaina LM, ed. Matters of Intelligence. Dordrecht, Holland: Reidel 1987; 59-87.
7. Maunsell JHR, Gibson JR. Visual response latencies in striate cortex of the macaque monkey. J Neurophysiol 1992; 68:1332-1344.
8. Oram MW, Perrett DI. Responses of anterior superior temporal polysensory (STPa) neurons to "biological motion" stimuli. J Cog Neurosci 1994; 6:99-116.
9. Raiguel SE, Lagae L, Gulyas B, Orban GA. Response latencies of visual cells in macaque areas V1, V2 and V5. Brain Res 1989; 493:155-159.
10. Nowak LG, Munk MHJ, Girard P, Bullier J. Visual latencies in areas V1 and V2 of the macaque monkey. Vis Neurosci 1995; 12:371-384.
11. Vogel R, Orban GA. Activity of Inferior temporal neurons during orientation discrimination with successively presented gratings. J Neurophysiol 1994; 71:1428-1451.
12. Perrett DI, Hietnan JK, Oram MW, Benton PJ. Organization and functions of cells responsive to faces in the temporal cortex. Phil Trans R Soc Lond 1992; 8 335:23-30.
13. Dinse HR, Kruger K. The timing of processing along the visual pathway in the cat. NeuroReport 1994; 5:893-897.
14. Allison T, Ginter H, McCarthy G, Nobre AC, Puce A, Luby M, Spencer DD. Face recognition in human extrastriate cortex. J Neurophysiol 1994; 71:821-825.
15. Thorpe S, Fize D, Mariot C. Speed of processing in the human visual system. Nature 1996; 381:520-522.
16. Treves A, Rolls ET, Tovée MJ. On the time required for recurrent processing in the brain. In: Torre V, Conti F, eds, Neurobiology: Ionic Channels Neurons and the Brain. New York: Plenum Press 1996; 371-382.
17. McClurkin JW, Optican L, Richmond BJ. Cortical feedback increases visual information transmitted by monkey parvocellular lateral geniculate-nucleus neurons. Vis Neurosci 1994; 11:601-617.
18. Ramachandran VS. 2-D or not 2-D—that is the question. In: Gregory RL, Harris J, eds. The Artful Eye. Oxford: Oxford University Press 1994; 249-267.
19. Tovée MJ, Rolls ET, Ramachandran VS. Rapid visual learning in the neurons of the primate temporal visual cortex. NeuroReport 1996; 7:2757-2760.

Index

S

T

V

W

MIX
Papier aus verantwortungsvollen Quellen
Paper from responsible sources
FSC® C105338

If you have any concerns about our products,
you can contact us on
ProductSafety@springernature.com

In case Publisher is established outside the EU,
the EU authorized representative is:
Springer Nature Customer Service Center GmbH
Europaplatz 3, 69115 Heidelberg, Germany

Printed by Libri Plureos GmbH
in Hamburg, Germany